MARINE STEAM ENGINEERING

MARINE STEAM ENGINEERING

Editor
William Dias

Published by:

DISCOVERY PUBLISHING HOUSE
4383/4B, Ansari Road, Darya Ganj
New Delhi-110 002 (India)
Phone : +91-11-23279245; 23253475; 43596065
Mobile : +91 9811179893 / +91 9871656464
E-mail : discoverybooksindia@gmail.com
orderdphbooks@gmail.com
namitwasan9@gmail.com
web : www.discoverypublishinggroup.com

Marine Steam Engineering
Edited by: **William Dias**

International Standard Book Number: **978-81-19365-82-1** (Hardback)

Printed at:
Infinity Imaging Systems
Delhi (INDIA)

PREFACE

The field of marine steam engineering has played a pivotal role in shaping the development and progress of maritime transportation. From the earliest days of steam-powered vessels to the present era of advanced marine propulsion systems, steam engineering has remained a vital component of ship design and operation. This comprehensive book aims to provide readers with a thorough overview of marine steam engineering, covering a wide range of topics and catering to individuals from diverse backgrounds. It offers an extensive exploration of the subject matter, encompassing both theoretical foundations and practical aspects, thereby making it suitable for a broad audience including students, researchers, engineers, and professionals involved in the maritime industry. It explores the early development of steam engines, their integration into ships, and the pivotal role they played in revolutionizing the maritime industry. Moving forward, the subsequent chapters of the book delve into the fundamental principles of marine steam engineering. It covers essential topics such as thermodynamics, heat transfer, fluid mechanics, and combustion, elucidating their applications in steam power generation and propulsion systems. The theoretical concepts are complemented by practical examples and case studies, bridging the gap between theory and real-world applications. This approach equips readers with the necessary tools to analyze and understand the complex systems at work.

The book also dedicates significant attention to the design, construction, and operation of marine steam boilers and engines. It explores various boiler designs, their working principles, and safety considerations. Additionally, it delves into the different types of steam engines utilized in marine vessels, including reciprocating engines, turbine engines, and combined cycle systems. The discussions encompass engine performance, efficiency optimization, and emission control strategies, all of which are crucial factors in modern marine steam engineering.

By presenting a holistic view of the field, the book serves as a valuable resource for readers seeking a comprehensive understanding of marine steam engineering and its future direction. It offers insights into the current state of the industry and highlights the emerging trends and technologies that will shape its trajectory.

The editor has collected reliable and scientifically proven information from various credible sources to provide readers with the latest and most accurate information available. The book is a compendium of contributions from various experts in the field, and the editor expresses gratitude to these contributors for sharing their knowledge and expertise.

Editor

CONTENTS

CHAPTER 1

Improvement of Marine Steam Turbine Conventional Exergy Analysis by Neural Network Application

Sandi Baressi Šegota[1], Ivan Lorencin[1], Nikola Anđelić[1], Vedran Mrzljak[2] and Zlatan Car[1]

[1]*Department of Automation and Electronics, Faculty of Engineering, University of Rijeka, Vukovarska 58, 51000 Rijeka, Croatia*
[2]*Department of Thermodynamics and Energy Engineering, Faculty of Engineering, University of Rijeka, Vukovarska 58, 51000 Rijeka, Croatia*

ABSTRACT

This article presented an improvement of marine steam turbine conventional exergy analysis by application of neural networks. The conventional exergy analysis requires numerous measurements in seven different turbine operating points at each load, while the intention of MLP (Multilayer Perceptron) neural network-based analysis was to investigate the possibilities for measurements reducing. At the same time, the accuracy and precision of the obtained results should be maintained. In MLP analysis, six separate models are trained. Due to a low number of instances within the data set, a 10-fold cross-validation algorithm is performed. The stated goal is achieved and the best solution suggests that MLP application enables reducing of measurements to only three turbine operating points. In the best solution, MLP model errors falling within the desired error ranges (Mean Relative Error) MRE $< 2.0\%$ and (Coefficient of Correlation) $R^2 > 0.95$ for the whole turbine and each of its cylinders.

Keywords: *exergy destruction; exergy efficiency; Marine steam turbine; MLP neural network; turbine cylinders*

1. INTRODUCTION

The dominant usage of steam turbines worldwide is related to electrical generator drive and electricity production [1,2]. Steam power plants, with steam turbines as essential components, can be assembled by following various methodologies.

Along with conventional steam power plants [3] and nuclear power plants [4,5], the novel approach in steam power plant design is the usage of various renewable energy sources which can notably improve steam power plant operation and its efficiency, and which are very beneficial to the environment [6,7]. The reduction of harmful emissions from such plants is today one of the most important research and scientific topic and multiple researchers are developing various techniques and processes with a goal of emissions reduction [8-10].

In addition to being used as stand-alone systems, steam power plants can be integrated into more complex systems, such as combined cycle power plants [11]. In combined cycle power plants, waste heat from the gas turbine is used for superheated steam production-in such a way, the environment is protected from huge waste heat amount and the reduction in harmful emissions (in comparison to pure gas or pure steam power plants) is also notable [12,13]. Such an operation of combined cycle power plants results in high efficiency, much higher in comparison to the conventional or nuclear steam power plants [14,15].

In marine power systems, internal combustion engines take a dominant share in the entire world fleet [16]. Due to internal combustion engine dominancy, various researchers are involved in investigating improvements as well as in minimizing the overall harmful impact on the environment [17-19]. Steam power plants are generally rarely used in marine power systems. However, there are several marine engineering fields in which steam power plants are still dominant. Additionally, steam power plants can be found as a part of new, complex marine systems that are currently under development [20,21]. All aforementioned benefits of combined cycle power plants are also present in the marine power systems [22,23]. The complexity of combined cycle power plants, especially for marine usage, requires adequate control and regulation systems for its proper (or if possible, optimal) operation.

One of the marine fields in which steam power plants are still predominantly used is the propulsion of LNG (Liquefied Natural Gas) carriers, but it should be noted that the utilization share of internal combustion engines, especially dual-fuel engines, is also increasing [24,25]. There are several steam turbines in the steam propulsion plant of any steam-powered LNG carrier. Along with the main turbine used for the propulsion propeller (or several propellers) drive, in such plants two or more steam turbines are mounted for the electrical generators drive (turbogenerators) [26,27] and low power steam turbine for the main feed water pump drive [28].

Analysis of any component from the steam power plant can be performed by using various approaches and techniques presented in the literature [29,30]. For the analysis of the main marine steam turbine observed in this research, exergy analysis is used. Exergy analysis of any component or the entire system is a technique that offers many benefits in comparison to other analysis methods. Exergy analysis does not take into consideration processes that occur inside any component-for the exergy analysis, only fluid flows and heat transfer (to and from the analyzed component) as well as used or produced mechanical power are necessary. Therefore, for the exergy analysis, details about the analyzed component's inner structure are not required, simplifying all necessary measurements [31,32]. On the other hand, a lack of information about the inner structure of the observed component does not allow research and analysis of many details and processes inside the component. All the exergy analysis benefits can be seen in a variety of scientific papers that take this analysis as the baseline.

If considering the analyses of the entire power plants, Ahmadi and Toghraie [33] applied exergy analysis for the investigation of Montazeri steam power plant in Iran, while Si et al. [34] used the same analysis for the investigation of a 1000 MW double reheat ultra-supercritical power plant. Ibrahim et al. [35] analyzed the thermal performance of the gas turbine power plant, while Aghbashlo et al. [36] observed the performance assessment of a wind power plant by using exergy analysis. AlZahrani and Dincer, I. [37] observed parabolic trough solar power plant and Abuelnuor et al. [38] investigated Garri "2" combined cycle power plant also by using exergy analysis.

Exergy analysis is successfully applied in the performance analysis of many components and processes from various plants. Zhao et al. [39] used exergy analysis for the investigation of the turbine system in a 1000 MW double reheat ultra-supercritical power plant. Medica-Viola et al. [40] used exergy analysis for the performance analysis of low-power steam turbine with one extraction used in marine applications. Presciutti et al. [41] applied exergy analysis for the investigation of glycerol combustion in an innovative flameless power plant. Szablowski et al. [42] used exergy analysis for the investigation of an adiabatic compressed air energy storage system. Arshad et al. [43] performed a review of the exergy analysis usage in the investigation of fuel cells. Lorencin et al. [44] applied exergy analysis for the analysis of steam mass flow rate leakage through steam turbine labyrinth (gland) seals.

Exergy analysis can also be a baseline for the economic analysis of various power plants or its components [45-47]. From the literature, it can be found that exergy analysis is used in the investigation and observation of many other plants, processes and components.

Along with exergy analysis, an extensive literature review also shows that many scientists and researchers used various artificial intelligence methods and processes in the analysis of power plants or its components.

One of the most used artificial intelligence methods in the energy sector is MLP (Multilayer Perceptron) neural network. Sun et al. [48] developed a new MLP-based soft sensor for SO_2 power plant emissions detection. Several researchers [49-52] used MLP for predicting electrical power output from various complex power plants. Wahid et al. [53] applied MLP for the prediction of energy consumption in the buildings. Tahan et al. [54] used MLP for condition-based maintenance of gas turbine, while Lorencin et al. [55] also used MLP for condition-based maintenance, but not for the gas turbine only, then for the entire marine CODLAG (Combined Diesel and Gas) propulsion system. Many other authors also used MLP for the condition-based maintenance problems of various plants and components [56,57].

MLP can also be used for predicting ship speed by using some of the ship propulsion system parameters [58]. Detecting and diagnosing faults by applying MLP in a steam turbine that operates in a thermal power plant was presented Dhini et al. [59], while Tian et al. [60] and Ayo-Imoru and Cilliers [61] used MPL for detecting various losses and prevention of accidents in the nuclear power plants. Various other neural network applications can also be found in the literature in many energy sectors and processes [62-64].

An extensive literature review shows that MLP is not used currently for tracking operating parameters or performances of the main marine steam turbine and all its cylinders. Additionally, the possibility that measurements are reduced by MLP neural network application is not investigated. During the possible reduction of measurements, the dominant goal for MLP must be high accuracy and precision in the prediction of any required operating parameter. The intention of this paper is not only to fill the literature gap but also to show possibilities that neural network applications offer in marine systems (or its components) and to be a guideline for other researchers interested in this field.

In the presented paper, exergy analysis of the main marine steam turbine (as well as both of its cylinders) is performed. The analysis is based on the measurement data obtained during steam turbine exploitation at 24 different loads. At the beginning, the conventional exergy analysis is performed, which requires many measurements at each turbine load. After conventional analysis, an exergy analysis is performed by the MLP neural network application. The application of MLP can significantly reduce the amount of performed measurements, while the accuracy and precision of the obtained exergy analysis parameters remain high, regardless of the observed load. This analysis can be a guideline for reducing control and measurement equipment inside the marine steam power plant, especially on new ships.

2. DESCRIPTION AND OPERATION PRINCIPLE OF THE ANALYZED MAIN MARINE STEAM TURBINE

The main marine steam turbine analyzed in this paper operates at the 100,450 tons (gross tonnage) commercial LNG carrier. A steam turbine is used for the LNG carrier propulsion. The maximum mechanical power that can be produced by the observed turbine is 29,420 kW, according to the manufacturer specifications [65]. The general scheme of the analyzed turbine, along with operating points required for the exergy analysis, is presented in Figure 1.

The main marine steam turbine is composed of two cylinders and these are High Pressure Cylinder (HPC) and Low Pressure Cylinder (LPC). Each marine steam propulsion system consists of two parallel operating steam generators which produce superheated steam and delivers the majority of cumulatively produced steam mass flow rate (it depends on the system load) to the HPC inlet [66]. HPC has one steam extraction used for steam delivery to various auxiliary steam plant processes-HFO (Heavy Fuel Oil) heater, BOG (Boil Off Gas) heater, water heater for the crew requirements, etc. Both steam generators in this marine steam plant simultaneously use HFO and BOG during operation. After extraction, the remaining steam mass flow rate expands through HPC until the cylinder outlet. HPC consists of one Curtis and seven Rateau stages.

The analyzed main steam turbine is an older variant of marine propulsion steam turbines and it is designed without steam reheating. Newer variants of

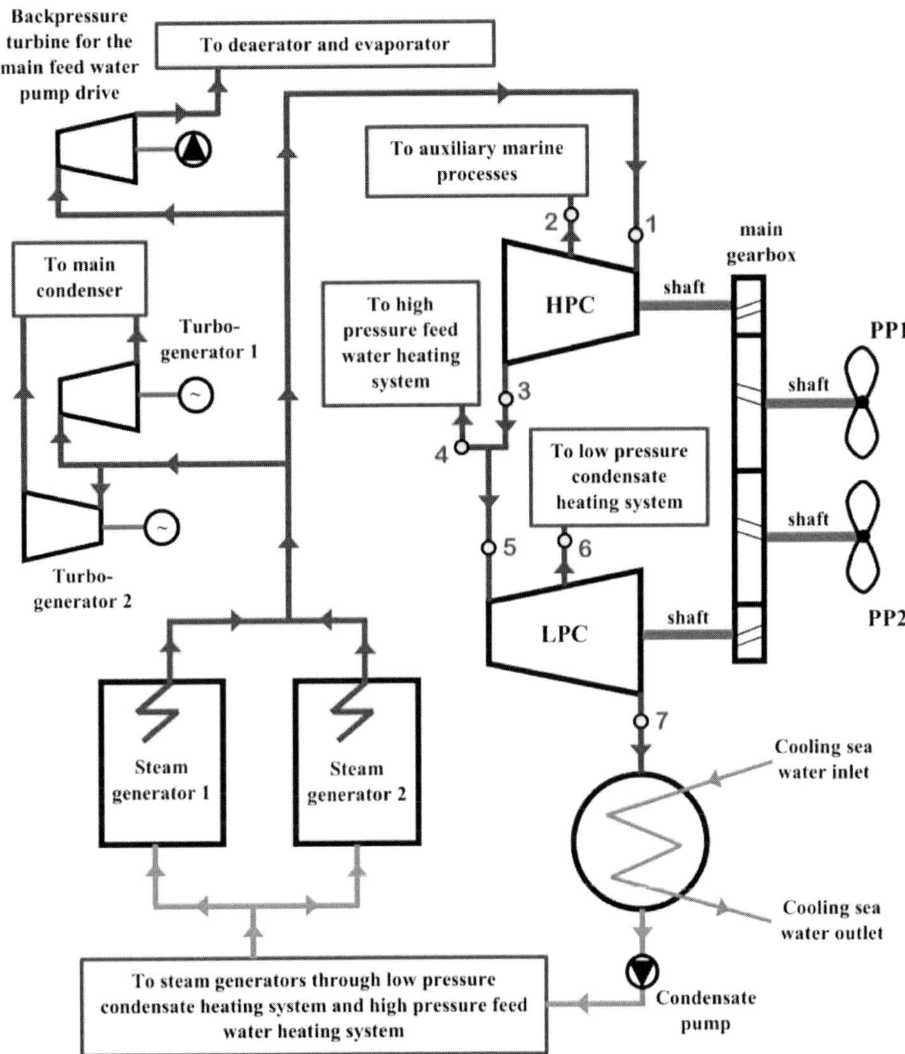

Figure 1. *Scheme of the main marine steam turbine along with operating points required for the exergy analysis. HPC: High Pressure Cylinder; LPC: Low Pressure Cylinder; PP1: The first Propulsion Propeller; PP2: The second Propulsion Propeller.*

marine propulsion steam turbines have one additional cylinder, Intermediate Pressure Cylinder (IPC), and steam reheating (steam reheaters are mounted inside steam generators) [67,68]. Such an upgrade increases plant overall efficiency, but simultaneously increases plant complexity and requires more stringent maintenance in comparison to older marine propulsion steam turbines.

Between the HPC and LPC of the analyzed turbine, one additional extraction is mounted for steam delivery to a high pressure feed water heating system. In the observed commercial LNG carrier, the high pressure feed water heating system consists of one high pressure feed water heater and deaerator [69]. In some operating regimes, when required, a part of steam extracted in this extraction (operating point 4, Figure 1), is delivered to air heaters used for heating of air at the steam generators entrance.

The remaining steam mass flow rate (operating point 5, Figure 1) expands through LPC. LPC, similar to HPC, has one steam extraction which is used for steam delivery to low pressure condensate heating system, which in the observed plant consists of one low pressure condensate heater and evaporator. An evaporator is a component used for the freshwater production (from sea water) and simultaneously for condensate heating [70]. After expansion in LPC, the remaining steam mass flow rate is delivered to the main marine steam condenser for condensation [71]. LPC consists of eight Rateau stages.

All three steam extractions from the observed turbine are not open all the time during the main turbine operation. Regulation valves opened and closed each of these extractions (and regulate extracted steam mass flow rate in each extraction) according to the predefined regulation procedure (following operation dynamic of the whole plant).

Both cylinders of the observed turbine are connected to the main marine gearbox through which one or two propulsion propellers are driven (in Figure 1 are shown two propulsion propellers, PP1 and PP2) [72].

It should be noted that in this exergy analysis (both conventional and with MLP application) several additional losses that occur in the plant, related to the main steam turbine, are neglected. For example, these losses are steam mass flow rate leakage through gland seals of each cylinder [73], heat losses in the pipelines and through the housing of each cylinder, mechanical losses [74], etc. Although important, all of these losses have a minor impact on the exergy analysis results of the observed steam turbine and each turbine cylinder.

3. CONVENTIONAL EXERGY ANALYSIS OF MAIN MARINE STEAM TURBINE AND EACH OF ITS CYLINDERS

Conventional exergy analysis of the main marine steam turbine and its cylinders is characterized by the fact that three steam operating parameters (steam temperature, pressure and mass flow rate) must be measured in each operating point presented in Figure 1 at each turbine load. Thus, each turbine load requires 21 measured data in order to be able to perform conventional exergy analysis at that particular load for the whole turbine and each cylinder. Change in turbine load requires new three measured data in each operating point. Therefore, conventional exergy analysis of observed main marine steam turbine and its cylinders require extensive measurements at each turbine load. Without these measurements (if any steam operating parameter, in any operating point from Figure 1 is missing), proper conventional exergy analysis cannot be performed or certain approximations must be used.

3.1. Overall Exergy Analysis Balances and Equations

In comparison to the energy analysis, of which results are not dependable on the ambient conditions [75,76], exergy analysis of any control volume or a system is dependable on the ambient conditions (ambient temperature and pressure) [77,78]. For the proper conventional exergy analysis of any control volume or a system, an overall exergy balance, mass flow rate balance and the most important variables should be defined. These overall equations and balances are valid in any exergy analysis, as well as in exergy analysis of the observed main marine steam turbine and both of its cylinders [79].

The overall steady-state exergy balance equation is defined as recommended in [80] by using an Equation (1):

$$\dot{Q}_{\mathrm{EX}} + P_{\mathrm{INLET}} + \sum Ex_{\dot{\mathrm{INLET}}} = P_{\mathrm{OUTLET}} + \sum \dot{\mathrm{E}}\mathrm{xOUTLET} + \dot{E}x_{\mathrm{DES}} \tag{1}$$

where P is the mechanical power (used or produced) and $\dot{E}x_{\mathrm{DES}}$ is exergy destruction (exergy loss). It should be highlighted that in the overall exergy balance equation, potential and kinetic energies are disregarded due to its low influence on the overall balance. For the analyzed main marine steam turbine and its cylinders potential and kinetic energies in above balance are also low, and their inclusion will not bring meaningful change in the obtained results [81]. $\dot{Q}_{\mathrm{EX}}$ is the exergy transfer by heat at the temperature T, of which the definition can be found in the literature [82] through the following Equation (2):

$$\dot{Q}_{\mathrm{EX}} = \sum \left(1 - \frac{T_0}{T}\right) \cdot \dot{Q} \tag{2}$$

where $\dot{Q}$ is an energy transfer by heat, T is temperature and index 0 corresponds to the state of the ambient. The last undefined variable from the overall exergy balance equation is a total exergy power of operating medium flow (E□x), of which definition can be found in [83], Equation (3):

$$\dot{E}x = \dot{m} \cdot \varepsilon \tag{3}$$

where $\dot{m}$ is operating medium mass flow rate and ε is specific flow exergy of operating medium. Operating medium specific flow exergy is calculated according to the Equation (4), [84]:

$$\varepsilon = (h - h_0) - T_0 \cdot (s - s_0)\,, \tag{4}$$

where h is operating medium specific enthalpy and s is operating medium specific entropy. During control volume or a system standard operation, mass flow rate leakage did not occur. By taking into account the fact that for the analyzed main marine steam turbine and its cylinders all the standard small steam leakages (as for example, leakage through gland seals of each cylinder) are neglected as described above, the valid mass flow rate balance is [85], Equation (5):

$$\sum \dot{m}_{\mathrm{INLET}} = \sum \dot{m}_{\mathrm{OUTLET}}. \quad (5)$$

The overall definition of the exergy efficiency can be presented as proposed in [86], Equation (6):

$$\eta_{\mathrm{EX}} = \frac{\text{CUMULATIVE EXERGY OUTLET}}{\text{CUMULATIVE EXERGY INLET}}, \quad (6)$$

with the note that exergy efficiency of any observed control volume or a system can significantly differ from the overall definition, which depends on operating characteristics and operation principles of each control volume or a system.

3.2. Equations for the Exergy Aanalysis of Main Marine Steam Turbine and Its Cylinders

For each cylinder and the whole turbine, the first step is the calculation of the developed mechanical power. This step represents the essential element in exergy analysis equations. After developing the mechanical power equations, equations for the calculation of exergy destruction and exergy efficiency will be presented for each cylinder and the whole turbine. All the equations are defined according to recommendations from the literature [87,88] and are related to operating points presented in Figure 1.

3.2.1. High Pressure Cylinder (HPC)

Developed mechanical power, Equation (7):

$$P_{\mathrm{HPC}} = \dot{m}_1 \cdot (h_1 - h_2) + (\dot{m}_1 - \dot{m}_2) \cdot (h_2 - h_3). \quad (7)$$

Exergy destruction (exergy loss), Equation (8):

$$\dot{E}x_{\mathrm{DES,HPC}} = \dot{E}x_1 - \dot{E}x_2 - \dot{E}x_3 - P_{\mathrm{HPC}}. \quad (8)$$

Exergy efficiency, Equation (9):

$$\eta_{\mathrm{EX,HPC}} = \frac{P_{\mathrm{HPC}}}{\dot{E}x_1 - \dot{E}x_2 - \dot{E}x_3}. \quad (9)$$

3.2.2. Low Pressure Cylinder (LPC)

Developed mechanical power, Equation (10):

$$P_{\mathrm{LPC}} = \dot{m}_5 \cdot (h_5 - h_6) + (\dot{m}_5 - \dot{m}_6) \cdot (h_6 - h_7). \quad (10)$$

Exergy destruction (exergy loss), Equation (11):

$$\dot{E}x_{\mathrm{DES,LPC}} = \dot{E}x_5 - \dot{E}x_6 - \dot{E}x_7 - P_{\mathrm{LPC}}. \quad (11)$$

Exergy efficiency, Equation (12):

$$\eta_{\mathrm{EX,LPC}} = \frac{P_{\mathrm{LPC}}}{\dot{E}x_5 - \dot{E}x_6 - \dot{E}x_7} \quad (12)$$

3.2.3. Whole Turbine (WT)

Developed mechanical power, Equation (13):

$$P_{WT} = P_{HPC} + P_{LPC} \tag{13}$$

Exergy destruction (exergy loss), Equation (14):

$$\dot{E}x_{DES,WT} = \dot{E}x_1 - \dot{E}x_2 - \dot{E}x_4 - \dot{E}x_6 - \dot{E}x_7 - P_{WT}. \tag{14}$$

Exergy efficiency, Equation (15):

$$\eta_{EX,WT} = \frac{P_{WT}}{\dot{E}x_1 - \dot{E}x_2 - \dot{E}x_4 - \dot{E}x_6 - \dot{E}x_7} \tag{15}$$

In the equations for the exergy destruction and exergy efficiency of the whole turbine and each cylinder, total exergy power of steam flow ($\dot{E}x$) and steam specific flow exergy (ε) should be calculated using Equations (3) and (4) in each operating point from Figure 1. Additionally, the overall exergy balance equation, Equation (1), and steam mass flow rate balance, Equation (5), should always be satisfied for each cylinder and the whole turbine at each load.

4. EXERGY ANALYSIS OF MAIN MARINE STEAM TURBINE AND EACH OF ITS CYLINDERS BY MLP NEURAL NETWORK APPLICATION

MLP is a neural network that consists of artificial neurons arranged into multiple layers. MLP consists of at least three layers — an input layer, output layer and one or more hidden layers [89,90]. Neurons in input layers are used to set inputs for the MLP model. The number of neurons in that layer is equal to the number of inputs of the data set, and their values are set to the number of input values contained within the data set [89, 91]. The subsequent layers consist of neurons whose values are calculated depending on inputs and connection weights-with each artificial neuron in the subsequent layer being connected to all the artificial neurons in the preceding neurons with weighted connections [92,93]. If it is assumed that the value of the neuron is y_i^k, where k represents the layer number and i the neuron number within the layer, then the value of that particular artificial neuron is calculated as the activated weighted sum of artificial neurons of the previous layer [89], Equation (16):

$$y_i^k = F\left(\sum_{j=0}^{n_k} \theta_{j,i}^{k-1} \cdot y_j^{k-1}\right), \tag{16}$$

where $\theta_{j,i}^{k-1}$ represents the weight of the connection between the artificial neuron j in layer $k-1$ $\left(y_j^{k-1}\right)$ and the artificial neuron i in the layer k $\left(y_i^k\right)$. F represents the activation function-the function used to map the value of neuron into the desired range of values. Commonly used activation functions may [94, 95] :

- Eliminate the unwanted values such as Rectified linear unit-ReLU ($y = \max(0, x)$)-used to eliminate negative values [95],
- Map the input files to a certain range such as sigmoid (logistic) function which maps the values to a range of $[0, 1]$ $\left(y = \frac{1}{1+e^{-x}}\right)$ or hyperbolic tangent function which maps them to the range of $[-1, 1](y = \tanh(x))$,
- Simply map the input directly to output as is the case with the identity activation function $(y = x)[96, 97]$.

As MLP belongs to the family of machine learning algorithms, it has the ability to adjust itself to the data used for training it. This is done through the process of training, divided into forward and backward propagation $[89, 98]$. In the forward propagation part of the training process, a single set of the input data values (steam temperature, pressure and mass flow rate in used operating points, Figure 1, along with the ambient pressure and temperature) are used as input neuron values (where the number of input neurons equals the total number of inputs). Then, weights of inter-neuron connections are set randomly and the values of neurons in the hidden layer. Finally, the output layer values are calculated using Equation (16) [89,92]. The output value is compared to the value of one of the outputs-either exergy destruction (exergy loss) or exergy efficiency-and the general output contained within the dataset is marked with y. It can be expected that the MLP output, marked $\hat{y}$, will have a certain error $\epsilon = \sqrt{(y - \hat{y})^2}$. This error is then used in the backward propagation process in order to adjust the weights based on the gradient of the error (with a higher error values causing a larger adjustment being made to the weights). If the vector of the weights in a layer k is marked as $\Theta^{\mathrm{k}} = \left[\theta_1^k \theta_2^k \cdots \theta_{n_k}^k\right]$, and learning rate with α, this can be written as [58,89], Equation (17):

$$\Theta^{\mathrm{k}}_{\mathrm{new}} = \Theta^{k}_{\mathrm{old}} - \alpha \frac{\partial \epsilon}{\partial \Theta^{\mathrm{k}}_{\mathrm{old}}} \tag{17}$$

By repeating the described training process for multiple sets of input and output values, the MLP weights can be finely adjusted and provide a very low error when used as a trained model. The dataset consists of 125 points, each of which has data entries for the ambient temperature and pressure, as well as steam mass flow rate, temperature and pressure in each of the seven operating points (Figure 1), as well as values of exergy destruction and exergy efficiency for high pressure cylinder, low pressure cylinder and the whole turbine. This means that each data point has 23 input values and 6 output values. It should be noted that by its nature, MLP can only regress a single value within a model, and the number of inputs need to be fixed. Due to this fact, each output and each separate input set need to have a separate model trained.

Each of the listed input combinations will have two models trained-one for each of the possible outputs-exergy destruction and exergy efficiency. Due to this, the total number of final models is 72. Each of the models trained will need to have hyperparameters adjusted to achieve a quality regression performance. Hyperparameters are values which describe the general architecture of the neural network used to train the model. Hyperparameters of the MLP need to be varied to achieve the best regression models for each case. One of the varied hyperparameters is the earlier described activation function of the hidden layer neurons [96]. Further hyperparameters include the number of hidden layers and the number of neurons per hidden layer expressed as $(k_1, k_2, \cdots k_n)$ in which the total number of layers is n and k_i represents the number of neurons in layer i. The algorithm used for calculating the weight values during the training process, called a solver, is also one of the varied hyperparameters [99]. Additional varied hyperparameters are the learning rate, which adjusts the rate of the weight adjustment during the backpropagation process, as well as the type of the learning rate-whether its value will remain constant or scale depending on the number of iterations [100]. Finally, hyperparameter is the L2 regularization parameter, which, if high, penalizes the inputs which have a high individual influence on the MLP output value-which can result in underfitted models [101].

In order to find the optimal set of hyperparameters, the Grid Search (GS) algorithm can be used. GS works in a way that it calculates all possible hyperparameter combinations. Then, a neural network is trained with each of this hyperparameter combinations [102]. In this manner, a wide range of hyperparameters can be tested. While the algorithm might not find the best possible combination of hyperparameters, with enough hyperparameters, it can find a hyperparameter combination which is close to the best one [103]. If needed, for example, if none of the yielded models provide satisfactory performance, possible hyperparameter values can be expanded or further refined, around the hyperparameter combination that provides the best results [58,104].

Finally, the metrics that will define the quality of the model solution need to be defined. In machine learning algorithms, models are evaluated by splitting the data set into training and testing portions [89, 98]. The training portion is used during the previously described training process, and the trained model is evaluated on the testing portion. This is done by performing solely the forward propagation of the training process in order to obtain pairs of predicted values $(\hat{y}_i)$ which can then be compared to the real values (y) from the data set. This will provide the value that describes the performance of the model. Two such metrics are used in this paper and those are coefficient of correlation (R^2) and mean absolute error (MAE).

R^2 defines the ratio of variance which exists inside the data set with the amount of variance contained in the results of the trained models [105]. Less unexplained variance means that the model is tracking the real data better, with higher values of R^2, which is defined in the range from 0 to 1 [106]. R^2 is defined by the Equation (18), [105]:

$$R^2 = 1 - \frac{S_{RESIDUAL}}{S_{TOTAL}} = 1 - \frac{\sum_{i=1}^{n}(y_i - \hat{y})^2}{\sum_{i=1}^{n}\left(y_i - \frac{1}{n}\sum_{i=1}^{n} y_i\right)^2}. \quad (18)$$

MAE provides a clearer, direct value of the error which using the MLP model introduces when used for regression [107]. MAE is defined as [107,108], Equation (19):

$$MAE = \frac{1}{n}\sum_{i=1}^{n} |y_i - \hat{y}_i| \tag{19}$$

As the data set used in this research is relatively small, the need for cross-validation arises. Cross-validation is a technique which allows a larger amount of data to be used for testing. As the splits of training and testing data are randomized, a situation can happen in which a bad model performs well on a randomized testing set — while its real performance on the entire dataset is comparatively low [50, 109, 110]. A K-fold cross-validation, with 10 folds, is performed. This technique is applied in the following manner: first, the data set is split into K splits. Then, the model with the architecture provided from the grid search is trained on the training set consisting of the mix of $K - 1$ subsets, with the remaining 1 being used as the testing set [92,111]. This process is performed K times, with no repetitions of testing splits-in other words, until all the splits have been used as the testing split exactly once. In this manner, the entire dataset is used as the testing set, providing more detailed information on the model performance. In the case of the 10 -fold K-fold cross-validation, each split is trained with 90% of the dataset (9 folds) used as the training set, and 10% used as the testing set (1 fold). The metrics defined with the Equations (18) and (19) are applied on all K training/testing set combinations, and the final scoring is expressed as the average score over the K fold, along with the standard error of that value. To summarize, for each of the 72 input/output combinations, 6144 different MLP model architectures are trained for 10 cross-validation folds, and evaluated using R^2 and MAE metrics across all folds. This provides the average and standard error values which allow the models for each parameter combination to be compared and the best achieved model hyperparameter values to be determined. This process is illustrated in Figure 2.

The implementation of the described algorithm is done in the Python programming language, using Scikit-Learn machine learning library. Scikit-Learn was chosen for this research because it implements all the necessary algorithms for the presented research. Namely, the described Multilayer Perceptron, as used for a regression problem, is implemented within the MLPRegressor function, which takes the hyperparameters as the input [99]. Grid Search and K-Fold cross validation were implemented using the GridSearchCV function, which takes the number of folds (10), the selected algorithm (MLPRegressor), possible hyperparameter values stored within a dictionary data structure (with the names of hyperparameters being used as keys) and the list of desired metrics as inputs [99, 100]. Helpfully, the metrics used are also implemented within Scikit-Learn, or to be more precise, within the metrics module as mean absolute error and R^2; both of which take the least actual dataset values and a list of predicted values as inputs [99, 101].

The training of the models was performed using University of Rijeka's Bura supercomputer. Models for each output and input combination were trained on

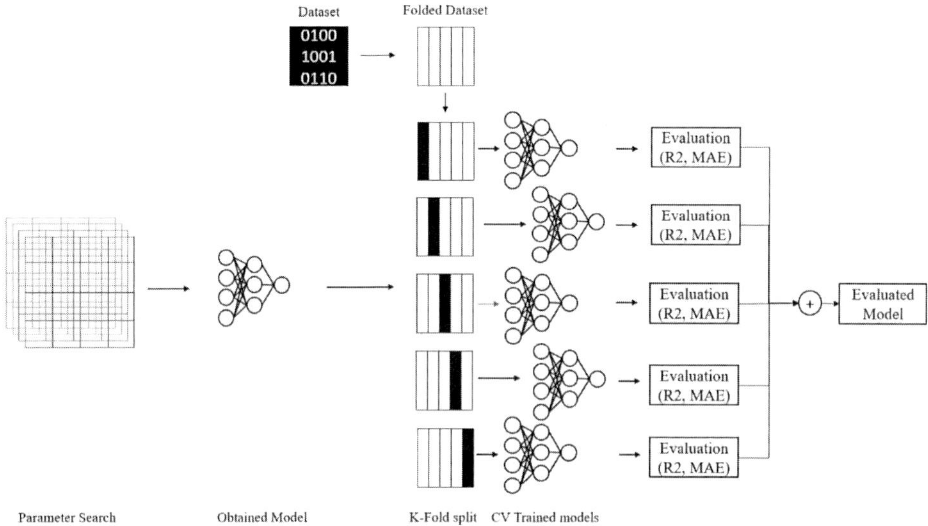

Figure 2. *The illustration of the Multilayer Perceptron (MLP) process used in the research, starting with the parameter search performed using a grid search (GS), yielding the MLP model, which is then applied on separate K-Fold splits and evaluated using R^2 and MAE metrics.*

a single node, so a total of 72 nodes were used. Each node of the Bura supercomputer consists of an Intel Xeon E5 CPU, which provides 24 physical or 48 logical cores, and 64 GB of RAM. At the time of the research Bura supercomputer used Red Hat Enterprise Linux operating system, with kernel version 3.10.0-957 [112]. Use of Scikit-Learn was enabled through Anaconda Data Science platform, version 4.8.4. [113].

5. STEAM OPERATING PARAMETERS REQUIRED FOR THE EXERGY ANALYSIS

For the purpose of conventional exergy analysis and exergy analysis by MLP neural network application of the main marine steam turbine and both its cylinders, in this research measurements are performed in each operating point from Figure 1 at 24 different turbine loads. Turbine load will be presented in relation to maximum turbine power (29, 420 kW) specified by the turbine manufacturer. The maximum measured load equals to 84.31% of maximum power, which corresponds with minimum specific fuel consumption of the steam propulsion plant. Measurements on the LNG carrier were obtained at 24 steady-state conditions. It should be highlighted that the authors did not have any permission to get involved in the ship operation or to influence the ship crew.

The conventional exergy analysis the complete data will be presented in three turbine loads-low, medium, and high load, which correspond to 7.03%, 49.79% and 83.22% of the maximum turbine power, respectively.

For the turbine exergy analysis by MLP neural network application, all the collected data are used at all measured turbine loads. Those data will not be fully

Table 1. *Steam operating parameters at low load (7.03% of maximum power)*

Operating Point *	Temperature (°C)	Pressure (MPa)	Mass Flow Rate (kg/h)
1	487	6.2	9622
2	-	-	0
3	235	0.097	9622
4	-	-	0
5	235	0.097	9622
6	-	-	0
7	62.13	0.00511	9622

* Operating point numeration refers to Figure 1.

Table 2. *Steam operating parameters at medium load (49.79% of maximum power).*

Operating Point *	Temperature (°C)	Pressure (MPa)	Mass Flow Rate (kg/h)
1	511	6.065	51,419
2	-	-	0
3	259	0.401	51,419
4	-	-	0
5	259	0.401	51,419
6	158	0.085	2985
7	28.85	0.00397	48,434

* Operating point numeration refers to Figure 1.

presented, for each measured steam operating parameter in each point from Figure 1 will be presented data range (from minimum to maximum) collected in all 24 turbine loads.

5.1. Conventional Exergy Analysis

Steam operating parameters in each operating point from Figure 1 at three different loads, required for the conventional exergy analysis of the main marine steam turbine and its cylinders are presented in Tables 1-3. In Table 1 steam data are presented for low turbine load, which corresponds to 7.03% of turbine maximum power, in Table 2 data are presented for medium turbine load (49.79% of maximum turbine power) and in Table 3 data are presented for high turbine load (83.22% of turbine maximum power).

Steam-specific enthalpy and specific entropy in each operating point at all loads are calculated from the measured steam temperature and pressure by using NIST-REFPROP 9.0 software [114]. Steam-specific flow exergy in each operating point for all turbine loads is calculated by using Equation (4). The steam specific flow exergy calculation requires definition of the ambient state in which analyzed steam turbine operates-in this research, the ambient state is defined as proposed in the literature [115] through the ambient temperature of 25°C and the ambient pressure of 1 bar.

By observing data from Tables 1-3, operation dynamics of the analyzed steam turbine can be seen during its load variations. At low turbine load (Table 1) all three steam extractions are closed-an increase in turbine load results in steam extractions

Table 3. *Steam operating parameters at high load (83.22% of maximum power).*

Operating Point *	Temperature (°C)	Pressure (MPa)	Mass Flow Rate (kg/h)
1	500	5.795	95,570
2	354	1.558	3398
3	250	0.590	92,172
4	250	0.590	13,172
5	250	0.590	79,000
6	154	0.120	4636
7	34.80	0.00557	74,364

* Operating point numeration refers to Figure 1.

opening, due to an increase in the steam mass flow rate delivered to the main turbine. The third and last steam extraction (operating point 6, Figure 1) is the first which will be open, through which a certain steam mass flow rate will be delivered to the components of low pressure condensate heating system (Table 2). A further increase in the turbine load results with opening of second steam extraction (operating point 4, Figure 1), and at high load follows the opening of first extraction. At the highest measured turbine loads, all three steam extractions will be opened, and through all of them, steam mass flow rate will be delivered to all required system components. Therefore, the first steam extraction (operating point 2, Figure 1) from the HPC is the last extraction which will be open at high turbine loads, Table 3.

Steam expansion process in the Mollier $h - s$diagram for all three main turbine loads in conventional exergy analysis are presented in Figure 3 [114]. Operating points numeration is performed according to Figure 1, while markings a, b and c denote 7.03%, 49.79% and 83.22% of turbine maximum power, respectively. From Figure 3 it can be seen that at low turbine load, steam after expansion (operating point 7a) is still superheated. In marine steam power systems at low load, in the steam flow stream at the main condenser entrance is injected certain amount of water. Injected water cool superheated steam and transferred it to the saturated state (changing of steam aggregate state in condenser at any load can be performed only if the steam is saturated). At higher loads (medium and high load), steam after expansion through the main turbine cylinders is saturated, and it did not require additional cooling (operating points 7*b* and 7c, Figure 3). It is also important to observe that an increase in turbine load shifts the whole expansion process closer to the saturation line. Such occurrence resulted with a fact that steam at the LPC outlet has higher content of water droplets as turbine load increases.

5.2. Exergy Analysis by MLP Neural Network Application

The inputs (operating points) in Table 4 have been selected based on the physical relation to each output and part (HPC, LPC and WT) that was modeled. Performing the modeling of all possible combinations of input parameters (operating points) for each desired output would significantly increase the computational complexity. It should also be noted that each operating point consists of measurements for steam temperature, pressure and mass flow rate. In addition to the steam

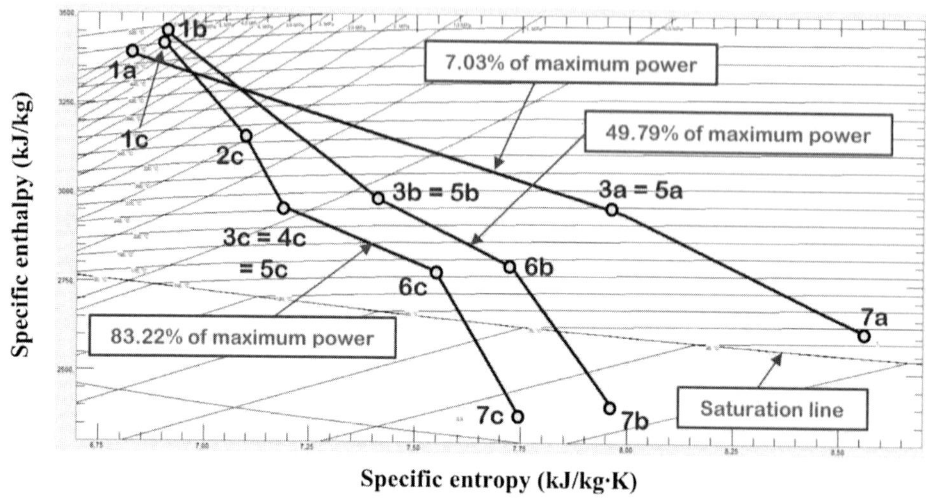

Figure 3. *Steam expansion process in Mollier h-s diagram for three observed turbine loads.*

temperature, pressure and mass flow rate in each of the listed points, each input set also includes the ambient temperature and pressure, with the first combination in all cases only including those two values.

Exergy analysis of the observed steam turbine and both its cylinders by using the MLP neural network is performed as follows:

1 By using all collected data, developed mechanical power, exergy destruction and exergy efficiency of each cylinder are calculated as well as the whole turbine at each of the 24 loads with the conventional exergy analysis.
2 Results obtained by conventional exergy analysis are then used for MLP training and testing.
3 MLP is trained for every hyperparameter combination given in Table 5, which results in a total of 442,368 models when the aforementioned cross-validation process is applied.
4 The results of 442,368 models are compared across 72 input/output parameter combinations, given in Table 5, in order to determine the best possible model architecture for each of the aforementioned combinations.

The results achieved with each trained MLP model are evaluated using the *MAE* and R^2 metrics described in previous section. This allows the determination of the best models achieved by the MLP method. This also allows the comparison between various operating points in order to determine the best possible measurement combination, or in other words, such a combination which will allow us to obtain the output value with as few operating points, under the condition that the metrics are satisfactory.

Steam operating parameters range (minimum-maximum) in each operating point of the observed main steam turbine (Figure 1) for all 24 measured turbine loads are presented in Table 6. From this table, it can be seen that steam temperature

Table 4. *Operating points used for regression models for each set of outputs. For each operating point, trained models use steam mass flow rate, temperature and pressure-in addition to ambient pressure and temperature.).*

Operating Points Combination	HPC * (Outputs: $\dot{E}x_{DES,HPC}$, $\eta_{EX,HPC}$)	LPC * (Outputs: $\dot{E}x_{DES,LPC}$, $\eta_{EX,LPC}$)	WT * (Outputs: $\dot{E}x_{DES,WT}$, $\eta_{EX,WT}$)
1	-	-	-
2	1,2,3,4,5,6,7	1,2,3,4,5,6,7	1,2,3,4,5,6,7
3	1,2,3,4,5	3,4,5,6,7	1,2,3,5
4	1,2,3	4,6,7	2,6,7
5	1,2	5,6,7	1,2,3
6	1,3	6,7	5,6,7
7	2,3	5,6	4,6,7
8	3,4	4,6	3,6,7
9	1,4	5,7	1,2,3,4,5
10	2,4	4,7	2,4,6,7
11	-	-	1,3,7
12	-	-	1,4,7
13	-	-	1,5,7
14	-	-	2,4,6
15	-	-	2,6,7
16	-	-	1,3,5,7
Count	10	10	16

* Operating point numeration refers to Figure 1.

and pressure at the HPC inlet (operating point 1) did not deviate significantly for the variety of turbine loads. This fact shows that both parallel operating steam generators delivers to the HPC steam with the highest temperature and pressure (specified by steam generators manufacturer) in the whole range of main turbine loads. Steam at the LPC outlet (operating point 7) can have high temperature at low turbine loads (which can reach up to 100°C, Table 6).

Figures 4 and 5 present the ranges of input parameters used in MLP training. Figure 4 shows the values of parameters used as inputs, with subfigure (a) showing the ranges of temperature, (b) the ranges of pressure and (c) the ranges of mass flow rate. The range of values is shown for each operating point, as indicated on the labels, along with the environment value range for temperature and pressure.

Figure 5 shows the range of output values, with subfigure (a) demonstrating the range of exergy efficiency values, while the subfigure (b) shows the exergy loss (exergy destruction) at each measurement point (HPC, LPC and WT).

It should be stated that exergy efficiency and exergy loss (exergy destruction) range starts from zero for the whole observed turbine and each of its cylinders. The reason of such occurrence is the first observed turbine load (of 24 overall loads). The first observed load is actually the heating of turbine (before start). In that load, the turbine and its cylinders did not produce useful power, so the exergy efficiencies and losses are equal to zero (losses are so small that it can be neglected).

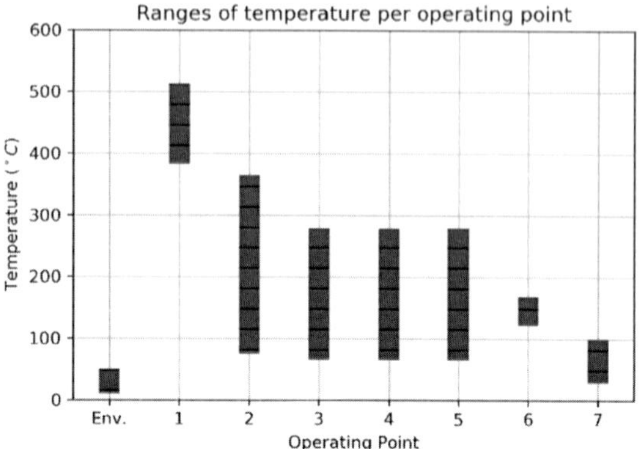

(a) **Ranges of temperature in each operating point.**

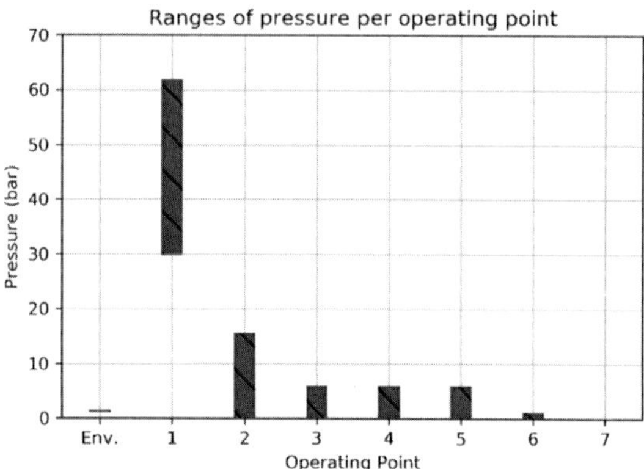

(b) **Ranges of pressure in each operating point.**

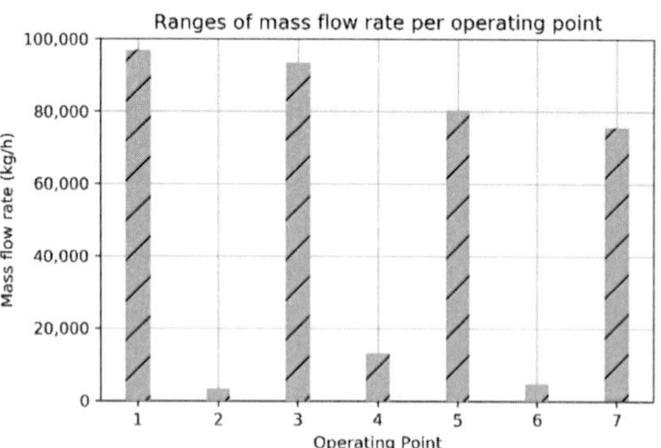

(c) **Ranges of mass flow rate in each operating point.**

Figure 4. *The ranges of measured values used as inputs into the MLP for (a) temperature, (b) pressure and (c) mass flow rate in each operating point, along with the environmental values for temperature and pressure (labeled as Env.).*

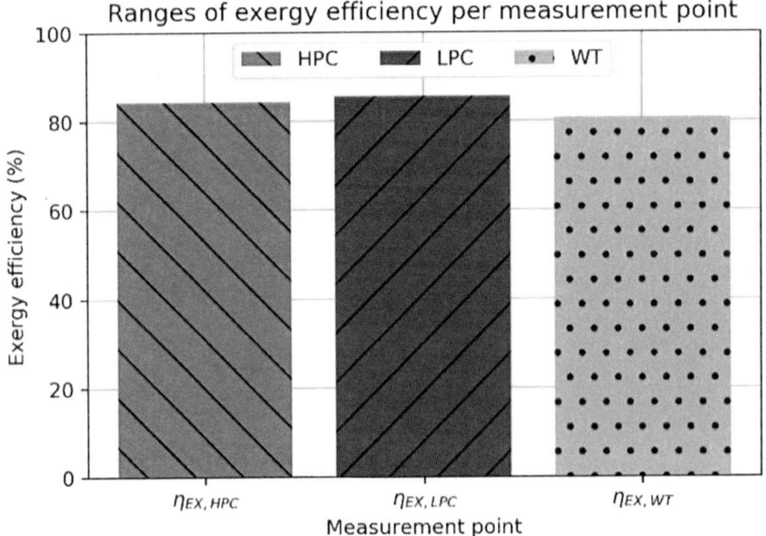

(a) Ranges of exergy efficiency in each output point.

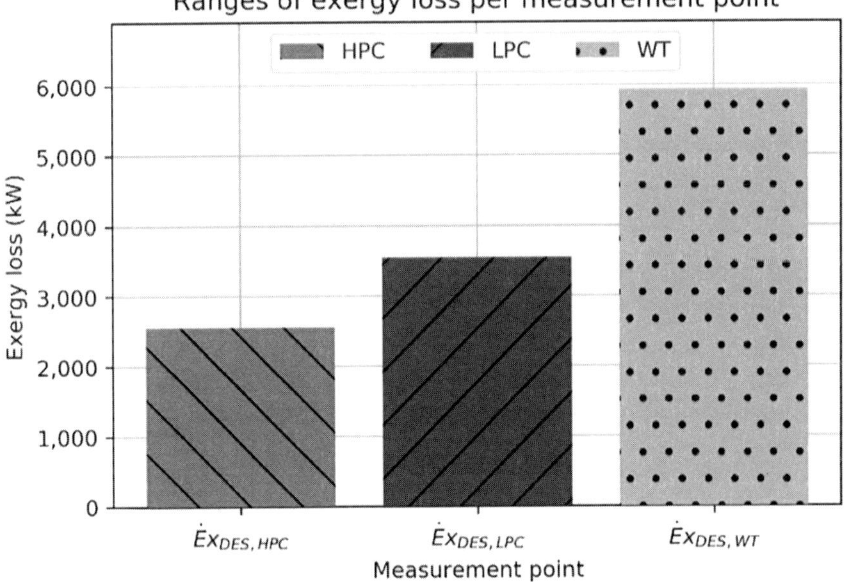

(b) Ranges of exergy loss in each output point.

Figure 5. *The range of measured values used as outputs into the MLP for (a) exergy efficiency and (b) exergy loss measured for HPC, LPC and WT.*

Table 5. *Possible values of hyperparameters used in grid search, with the number of the possible hyperparameter values being given in the column titled Total Count.*

Hyperparameter	Possible Hyperparameter Values	Total Count
Hidden Layer Sizes	(84,84,84,84) (84,84,84) (84,84) (84) (42,42,42,42) (42,42,42) (42,42) (42) (21,21,21,21) (21,21,21) (21,21) (21) (84,42,42,21) (42,21,21) (84,42,21) (42,21)	16
Activation Function	'relu' 'identity' 'logistic' 'tanh'	4
Solver	'adam' 'lbfgs'	2
Learning Rate Type	'constant' 'adaptive' 'inverse scaling'	3
Initial Learning Rate Value	0.5 0.1 0.01 0.00001	4
L2 Regularization parameter	0.1 0.01 0.001 0.0001	4

5.3. Measuring Equipment

Measurements of steam temperature, pressure and mass flow rate in each operating point from Figure 1 are performed with a standard measuring equipment, calibrated and already mounted inside the power plant. That measuring equipment is used for the main steam turbine process control and regulation during exploitation. The list of used measuring equipment is presented in Table 7 , while the detail specification of each measuring device can be found on the manufacturer

Table 6. *Steam operating parameters range (min-max) in each operating point for all 24 measured turbine loads.*

Operating Point *	Temperature (°C)	Pressure (MPa)	Mass Flow Rate (kg/h)
1	485–513	5.795–6.2	3835–96,789
2	283–365	0.08–1.565	0–3398
3	229–279	0.048–0.593	3835–93,521
4	229–279	0.048–0.593	0–13,202
5	229–279	0.048–0.593	3835–80,319
6	121–169	0.009–0.121	0–4772
7	28.616–100.02	0.00392–0.00561	3835–75,547

* Operating point numeration refers to Figure 1.

Table 7. *Used measuring equipment.*

<table>
<tr><th>Operating Point *</th><th>Temperature (Immersion Probes) [116]</th><th>Pressure (Pressure Transmitters) [117]</th><th>Mass Flow Rate (Differential Pressure Transmitters) [118]</th></tr>
<tr><td>1</td><td rowspan="2">Greisinger GTF 601-Pt100</td><td>Yamatake JTG960A</td><td rowspan="2">Yamatake JTD960A</td></tr>
<tr><td>2</td><td rowspan="6">Yamatake JTG940A</td></tr>
<tr><td>3</td><td rowspan="5">Greisinger GTF 401-Pt100</td><td rowspan="3">Yamatake JTD930A</td></tr>
<tr><td>4</td></tr>
<tr><td>5</td></tr>
<tr><td>6</td><td>Yamatake JTD920A</td></tr>
<tr><td>7</td><td>Yamatake JTD910A</td></tr>
</table>

* Operating point numeration refers to Figure 1.

website (provided in the list of references). Details related to measurement equipment accuracy and range of operation can be found in the Appendix A at the paper end. The measurement error did not have a significant influence on the obtained exergy analysis results.

6. RESULTS AND DISCUSSION

This section will present the results obtained first by the conventional exergy analysis, followed by the results obtained by the application of described AI methodology.

6.1. The Results of the Conventional Exergy Analysis

In the conventional exergy analysis of any steam turbine, from measured steam operating parameters firstly should be calculated produced mechanical power. For the case of this particular marine steam turbine, produced mechanical power is calculated not only for the whole turbine, but also for both turbine cylinders (HPC and LPC), see Figure 6. An increase in turbine load is followed by the increase in developed mechanical power of the whole turbine. However, from Figure 6, the share of each turbine cylinder in the cumulative developed mechanical power is interesting and important to observe. At low load, the dominant mechanical power

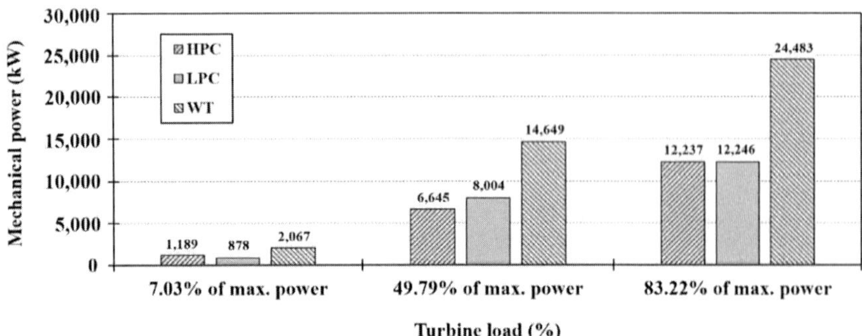

Figure 6. *Mechanical power developed by the main turbine and each of its cylinders at three observed loads.*

producer is HPC. An increase in turbine load results in a change in cylinder developed mechanical power share-at middle load the dominant mechanical power producer is LPC. At the highest measured loads, the share of both turbine cylinders in cumulative developed mechanical power is approximately the same; therefore, at the highest loads each cylinder develops approximately 50% of cumulative power. The same general conclusion about this type of marine steam turbines related to cylinder share in cumulative developed mechanical power at various loads can be found in the literature [66].

For three observed turbine loads in the conventional exergy analysis, developed mechanical power of HPC and LPC increases from 1189 kW and 878 kW at low load, to 6645 kW and 8004 kW at middle load and finally to 12, 237 kW and 12, 246 kW at high turbine load, respectively (Figure 6). It should be noted that developed mechanical power at each load presented in Figure 6 and calculated in each of 24 turbine loads is mechanical power calculated according to measured steam operating parameters and real (polytropic) steam expansion process throughout each cylinder. According to Figure 1, the mechanical power used for propulsion propellers drive in each turbine load is lower than the mechanical power calculated and presented in this analysis. The reason of such a difference is, as mentioned earlier, neglecting mechanical and other losses in the bearings, shafts and main gearbox.

Conventional exergy analysis at three observed turbine loads results with exergy destruction (exergy loss) and exergy efficiency of the whole main steam turbine and both of its cylinders, Figure 7. Comparison of Figures 6 and 7 shows that the developed mechanical power and exergy destruction of each cylinder and the whole turbine are directly proportional-higher developed mechanical power results in higher exergy destruction and vice versa. It is interesting to note that at a high load (83.22% of maximum power), regardless of the low difference between HPC and LPC developed mechanical power, LPC exergy destruction is significantly higher in comparison to HPC. Exergy destruction of the whole turbine increases during the load increase (proportional to developed mechanical power)-from 1380.09 kW at the low load, to 4437.86 kW at the middle load and finally to 5814.43 kW at the high load.

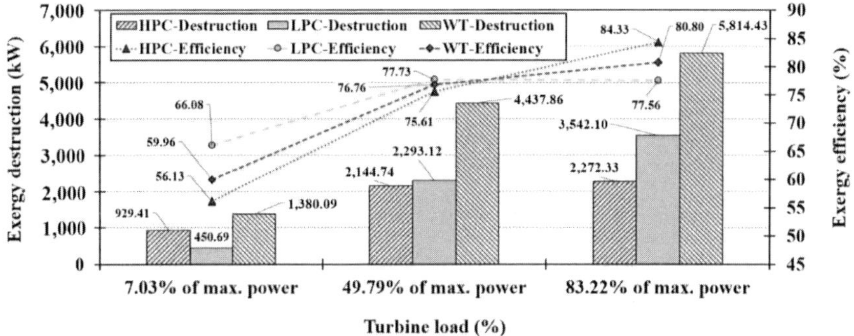

Figure 7. *Exergy destruction and exergy efficiency of the main turbine and each of its cylinders at three observed loads.*

Exergy efficiency results at three observed turbine loads in conventional exergy analysis show that an increase in turbine load increases the exergy efficiency of the whole turbine and each of its cylinders. The only deviation from this statement can be seen for the LPC, of which exergy efficiency slowly decreases from middle to high load (from 77.73% to 77.56%). Such a trend for LPC cannot be taken as relevant because the difference in exergy efficiency between middle and high load is so small that it can be the result of measurement equipment accuracy. By observing turbine cylinders, for low and high load a reverse proportionality between exergy destruction and exergy efficiency is valid-higher cylinder exergy destruction results in lower exergy efficiency. The mentioned reverse proportionality is not valid for the middle turbine load, where LPC, which has higher exergy destruction than HPC, also has higher exergy efficiency. An increase in the whole turbine load results in a simultaneous increase in its exergy efficiency-from 59.96% at a low load, followed by 76.76% at a middle load, to 80.80% at a high load (Figure 7).

In conclusion to previously described observations can be stated that the main marine steam turbine operation should be maintained at a high load where the whole turbine develops high mechanical power and has the highest exergy efficiency. Simultaneously, at a high load it should be taken into consideration that the whole turbine exergy destruction will be the highest, in comparison to lower loads. At a high load, both turbine cylinders will almost equally participate in cumulative developed mechanical power, but the exergy efficiency of HPC will be higher and its exergy destruction will be lower in comparison to LPC.

6.2. Exergy Analysis Results by MLP Neural Network Application

The results of MLP analysis are presented below. In order, the results for HPC, LPC and WT exergy destruction and efficiency are given. The graphs present the best results achieved using MLP per each given input combination, for each of the separate goals. In addition to *MAE*, Mean Relative Error (MRE) is used for presentation in order to demonstrate the error as a percentage of the range of the measured output. The results are not given for the case in which no input operating

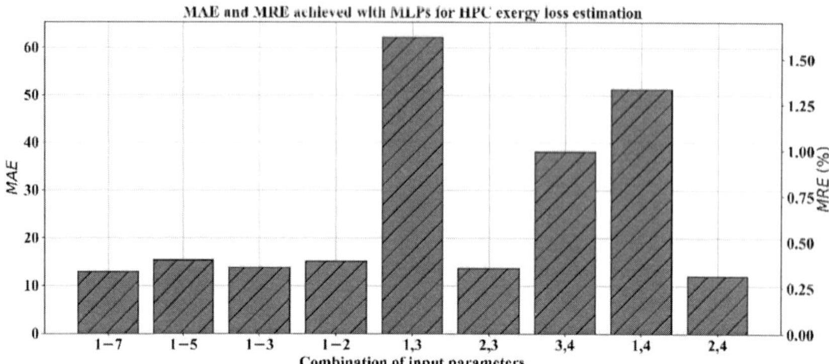

Figure 8. *MAE and MRE values for HPC exergy destruction (exergy loss).*

points have been used, using only ambient pressure and temperature, as the models for all possible hyperparameter combinations in all possible cases have failed to converge to a solution and as such have not provided any viable or meaningful results. The failure to converge in such a case suggests that such a regression, using only the ambient values, is insufficient for the desired outputs due to low or non-existent correlation between the ambient values and the desired outputs.

As for the acceptable error range, any MRE value that is smaller than 2% of the output range for a given output is considered to be acceptable, as this is precise enough estimation for the practical purposes in determining the exergy destruction and efficiency of a turbine in the marine environment. In the same manner, any R^2 value that is higher than 0.95 is considered within the acceptable range. This means that all models obtained with a certain input combination that have achieved an R^2 value higher than 0.95 and an MRE lower than 2% are to be considered when the final model selection is performed.

For all figures in this subsection, numbers or ranges written on the abscissa (combination of input parameters) are related to operating points presented in Figure 1.

MAE and MRE scores in the case of HPC exergy destruction (exergy loss) estimation display low errors across all inputs ($MRE < 2\%$), but there are some outliers. Namely, input combinations lacking the operating point $2(3,4;1,4;1,3)$ display the highest error, suggesting that operating point 2 is necessary to achieve a low model error in this case. This is shown in Figure 8.

Figure 9 shows the R^2 scores achieved being high across all input combinations, with all of them achieving scores in excess of 0.99. These suggest that all the input combinations may be used for modeling, since all models track the outputs well, as long as a higher error of some models does not present an issue. Still, it is evident that those combinations of operating points which do not include the operating point 2 achieve a lower R^2 score, which corresponds with the larger errors provided by those models. This further shows the importance of the operating point 2 in the modeling of HPC exergy loss-if higher precision than the one sought in this document is needed.

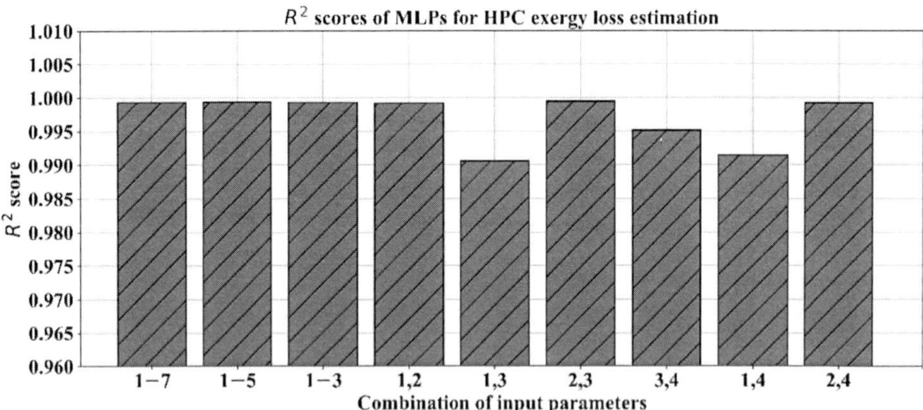

Figure 9. *R^2 values for HPC exergy destruction (exergy loss).*

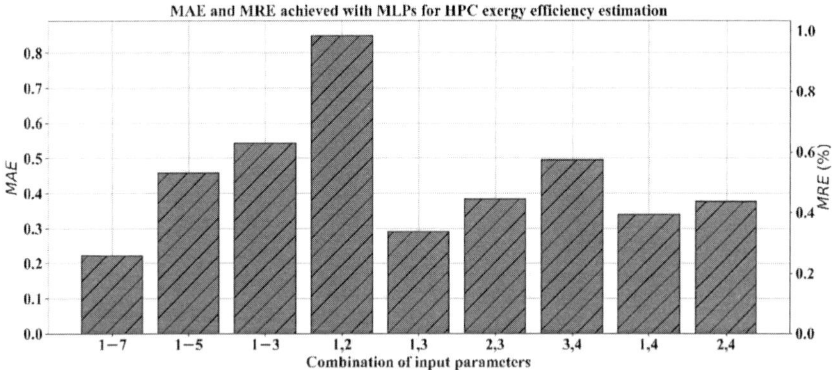

Figure 10. *MAE and MRE values for HPC exergy efficiency.*

While observing the MAE and MRE for HPC exergy efficiency estimation, given in Figure 10, it can be seen that all the input combinations achieve very low errors, namely below 1%. Interestingly, operating point 2, which seemed to have a large benefit, does not have as much importance. It seems that this role is replaced by the operating points 3 and 4 used in all combinations that achieve lower errors. As with the previous models, all the models here achieve MAE scores low enough to be considered for use.

Figure 11 shows that R^2 scores achieved for HPC exergy efficiency models are high, except in the case of input combination 1,2. Due to this combination being the only one which does not include operating points 3 or 4 , it suggests the importance of these inputs in HPC exergy efficiency model as the previous graph. In line with the MAE scores, the R^2 scores achieved are high enough $\left(R^2 > 0.95\right)$ to be considered usable in modeling HPC exergy efficiency.

While all the input combinations provide satisfactory metric values in terms of $MRE(MRE < 2\%)$ for HPC exergy destruction and exergy efficiency estimators, it

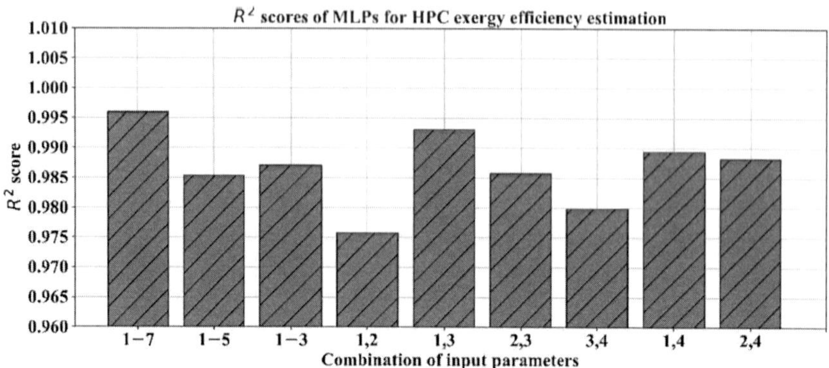

Figure 11. *R^2 values for HPC exergy efficiency.*

can be seen that there are some that do not achieve as high an R^2 value for exergy efficiency estimation $\left(R^2 > 0.99\right)$. Namely, only input combinations which achieve R^2 scores 0.99 or higher are $1-7; 1,3$ and 1,4 . It can also be seen that R^2 scores of those operating point combinations which achieve a higher error are lower, which is to be expected.

By observing all the achieved scores for the output models of HPC it can be shown that the best scores are achieved, for both exergy efficiency and exergy destruction, when the input operating point combinations used are 1,4;2,4;2,3 and 1-7. These operating point combinations are obtained by considering both errors and R^2 value scores for both exergy efficiency and exergy loss, and while they may not present the best possible model for an individual output, a high enough performance is evidenced in both observed output cases for HPC.

Figure 12 demonstrates the best MAE and MRE achieved by the MLP for each input combination in the case of exergy destruction estimation for LPC. It can be seen that the best results are achieved when all the inputs are used (1-7), with comparable results being achieved for all the input combinations, which include operating points 5, 6 and 7. Still, all the operating point combinations provide a satisfactory error below 1.5%, signifying that they may all be used in modeling the LPC exergy loss.

While observing the R^2 scores for LPC exergy destruction estimation presented in Figure 13 , it can be seen that all input combinations achieve relatively high R^2 scores $\left(R^2 > 0.95\right)$, meaning that all the models may be used in the LPC exergy destruction modeling. It should be noted that the input combination 5-7, despite a relatively low MAE in comparison to other results, has the lowest R^2 score. This points to the fact that, despite achieving a low error, this input combination does not explain all the variations in the test data. It can be concluded that this is due to the lack of information being contained within this input combination. Through comparison with the input combination of operating points 5 and 7 , which has achieved a higher score, it can be concluded that the inclusion of operating point 6 as an input is actually detrimental to the model, lowering instead of increasing performance.

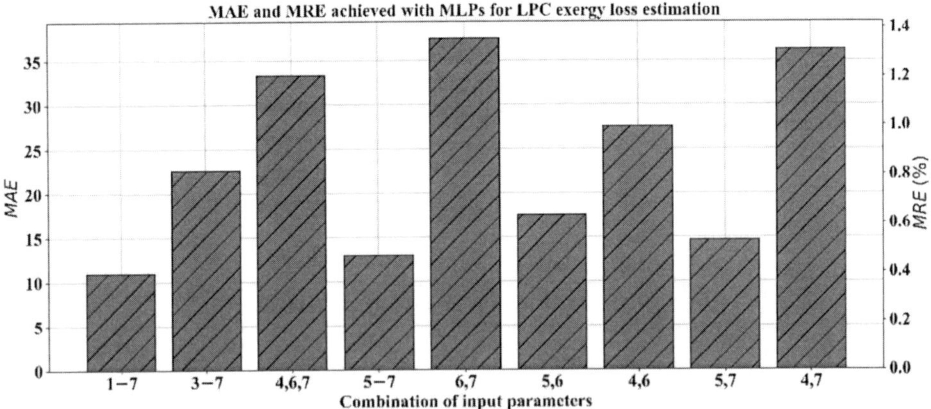

Figure 12. *MAE and MRE values for LPC exergy destruction (exergy loss).*

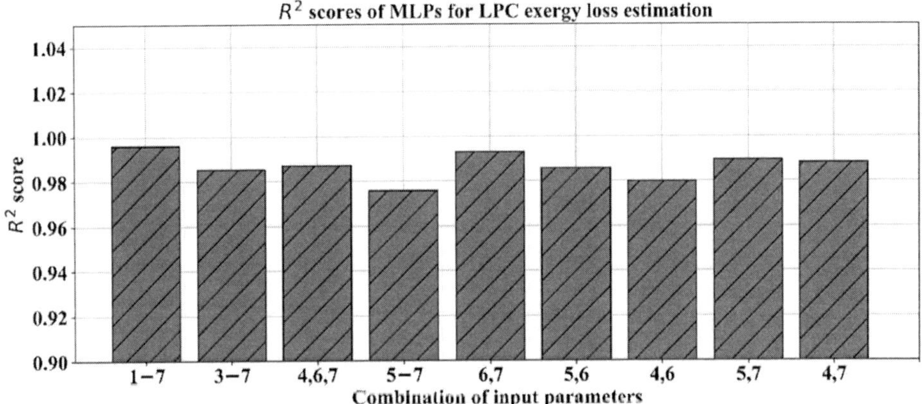

Figure 13. *R^2 values for LPC exergy destruction (exergy loss).*

For the model of LPC exergy efficiency, as in the previous case the best results are achieved when all inputs are used together (case 1-7), with comparable results being achieved for input combination 3-7. Despite this, all input combinations have achieved a low error ($MRE < 1\%$), with the highest error occurring when input combination 6,7 is used. Interestingly, the input combination 5-7 shows a lower error than combination 5,7, meaning that the inclusion of operating point 6 is beneficial in this case, but the use of the measurements in operating point 5 is largely beneficial to it. Because of this, we can conclude that the combination of values in operating points 5 and 6 is important to achieve extremely low scores in the LPC exergy efficiency modeling. However, as with the previous models, all the error values are low enough to conclude that all models may be used. These results are shown in Figure 14.

As for the R^2 scores of the model for LPC exergy efficiency in Figure 15, it can be seen that the lowest score is achieved by the input combination 6,7, which

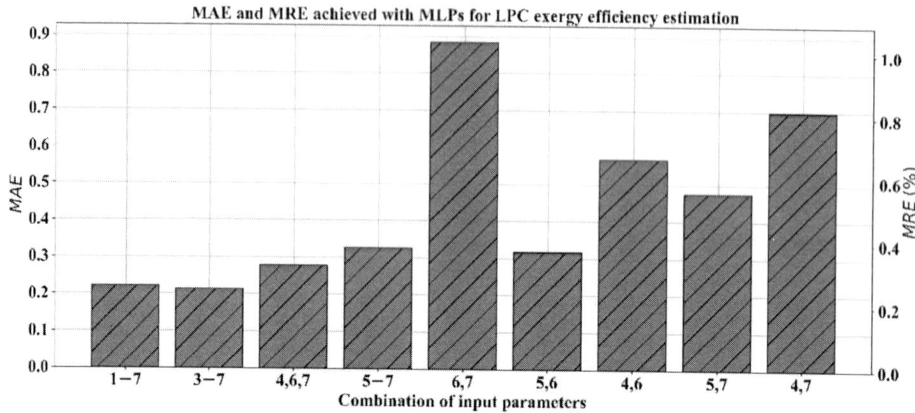

Figure 14. *MAE and MRE values for LPC exergy efficiency.*

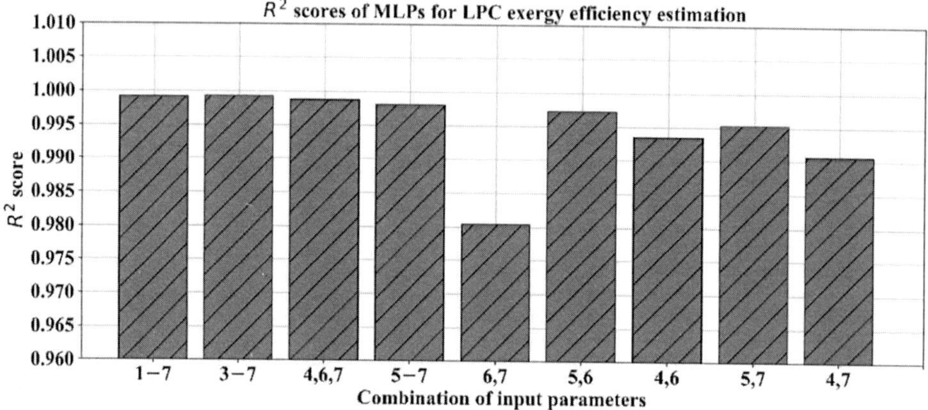

Figure 15. *R^2 values for LPC exergy efficiency.*

has the highest error (as can be seen in Figure 14). Still, this combination, and all others, provide a quality enough model of LPC exergy efficiency. This points to the lack of information necessary for a higher quality model when these inputs are used. Higher quality models being achieved in all other cases, which points to the fact that LPC exergy efficiency models require the use of operating point 4 or 5 , in order to achieve higher regression quality, and the scores achieved show that operating points 5 and 6 -when used in unison-provide enough information for MLP to successfully converge to a high-quality solution.

When observing the LPC exergy destruction and exergy efficiency estimation outputs (Figures 12-15), it can be seen that the best results are achieved when all the inputs are used, but satisfactory results are achieved for all input combinations with R^2 values in excess 0.97 , and MRE below 1.5%. Observing all the scores points towards the fact that the best results for the LPC outputs, including both exergy destruction and exergy efficiency, are obtained when points used are 1-7;5-7; 4, 6, 7 and 4,7.

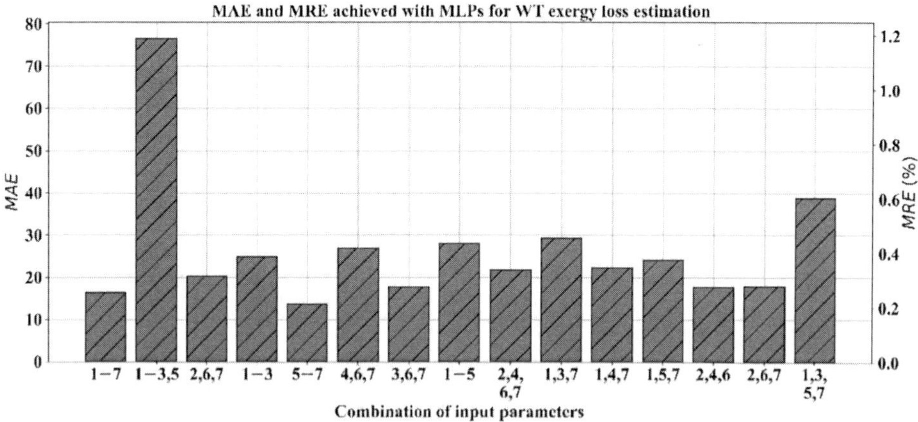

Figure 16. *MAE and MRE values for WT exergy destruction (exergy loss).*

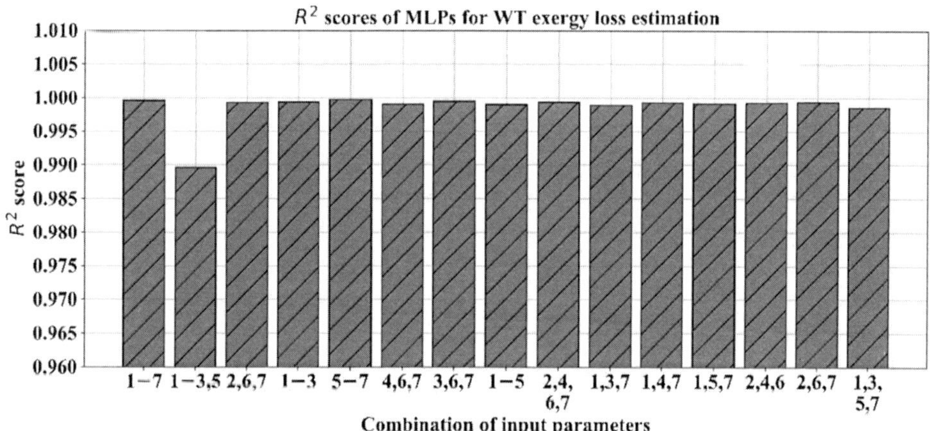

Figure 17. *R^2 values for WT exergy destruction (exergy loss).*

Figure 16 demonstrates *MAE* and *MRE* values obtained by the models for WT exergy loss estimation. It can be seen that all the error values are extremely low. The only input combination which exceeds the error of 1% is the combination of operating points 1-3,5, which achieves an error below 1.2%. However, such an error is still within the acceptable error range, meaning that all the models of WT exergy loss achieve satisfactory performance in regards to the *MAE*.

Figure 17 demonstrates the R^2 scores achieved by the WT exergy loss estimation fall in line with the *MAE* and *MRE* results achieved, with all the inputs achieving R^2 scores in excess of 0.985 ; meaning all are high enough to be considered for modeling. The lowest score is achieved by the input combination 1-3,5 which, with the value of 0.99, is still inside the acceptable range.

Observing both scores for WT exergy loss, it can be concluded that all models achieve similar high performance, with the exclusion of operating points 1,3,5,7

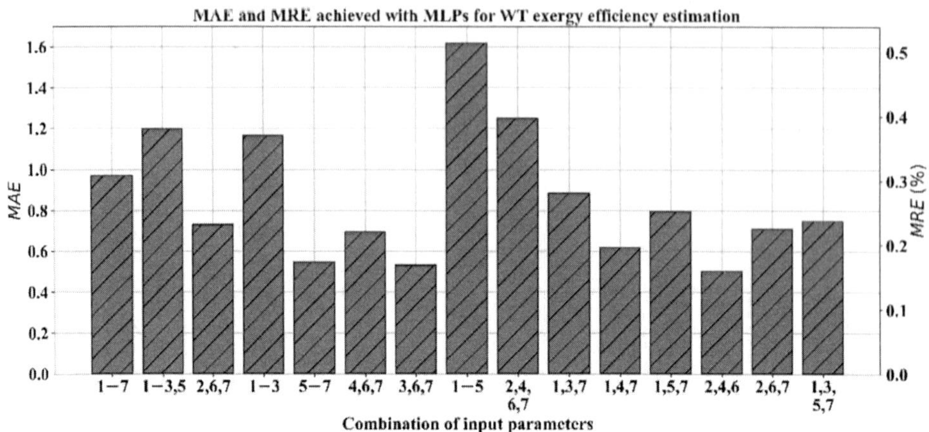

Figure 18. *MAE and MRE values for WT exergy efficiency.*

and 1-3,5, which achieve comparatively poorer scores. Still, all the model's scores fall well within the satisfactory ranges, indicating that it is possible to use them for the given task.

Through observing Figure 18, MAE and MRE of the WT exergy efficiency estimation can be seen. The output combination 1-5 achieves the lowest results, which is, interestingly, higher than the input combinations of $1-3,5$ and $1,3,5,7$, which do not include the operating point 4 . This signifies that operating point 4 may be detrimental in this particular case, but not so much that its inclusion should be avoided, considering the error introduced by it is at its maximum below 0.3%. Due to all MRE values being below 0.5 , this should not have a large influence on model selection.

Figure 19 shows the R^2 values achieved by the models for WT exergy efficiency estimation. The figure shows that the R^2 scores achieved mostly fall into line with the MAE achieved, with those models that show a higher error, also showing a lower R^2 score, but with less drastic differences. It can be seen that all the scores are in the excess of 0.99, which confirms that all the models are of high quality and that models with all the input combinations may be used during the model selection.

In the case of the output analysis of the whole turbine, it can be seen that error values are extremely low, except for the model of exergy destruction when the input combination of operating points 1-3,5 is used. When the input of $1-3,5$ operating points is used, the R^2 value drops below 0.99 and MRE grows above 1%. While these values are still within the limits of satisfactory results it can be concluded that they present the worst input combination in term of WT exergy destruction modeling, but may still be used if necessary. The best results achieved for both WT exergy destruction and efficiency are obtained when the following input parameters are used: $1-7; 1,4,7; 5-7; 2,4,6$.

Considering that all the used operating point combinations achieve MAE of below 2% for the output value range and R^2 score higher than 0.95 , all may be used for modeling the required outputs. For example, if a situation is observed in which

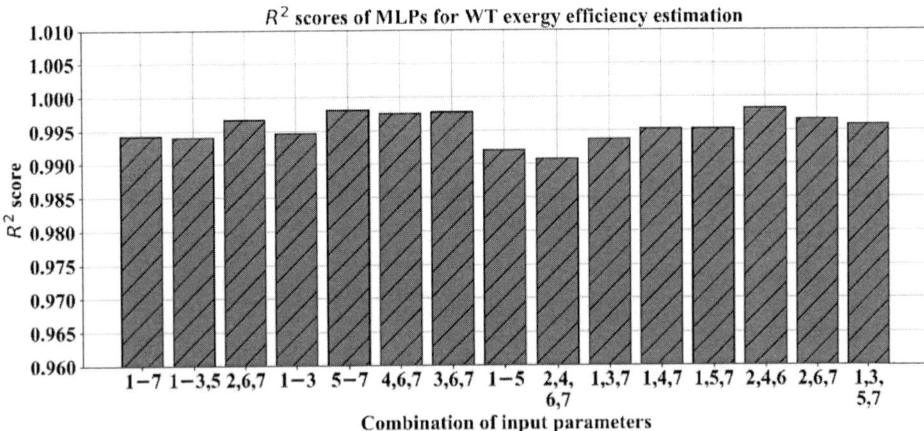

Figure 19. *R^2 values for WT exergy efficiency.*

the measuring equipment may already exist at the given points and installing it in different ones may pose a difficulty, using a model with slightly poorer scores (but still within the satisfactory range) can be a good choice. All the input operating point combinations which achieve quality models show the importance of including domain experts in the artificial intelligence-based research. In the presented research, this allowed for lowering the number of input combinations that were tested and consequentially for a lower computational complexity-while still generating useable models.

By observing the presented graphs, the best input combination can be determined. This can be done by cross referencing the best scores achieved in order to find the combination of inputs that provides the best scores. For example, while the best scores across all the observed cases are obtained when input combinations 1-7 are used, this does not lower the amount of operating points needed. Due to all of the output metrics falling well within the satisfactory error range, the selection can concentrate on finding such a combination of input operating points, which allows for an as low as possible number of operating points. If the goal is to achieve satisfactory measurements with as few operating points as possible, then it can be concluded that this combination is 1, 4 and 7-or namely by utilizing operating points 1 and 4 for HPC, 4 and 7 for LPC and 1,4,7 for WT. The respective values achieved for each cylinder and the whole turbine are given in the graphs below.

Selected operating points (1, 4 and 7) reduce the number of the required measurements for more than half (from 21 to 9 overall measurements of the steam mass flow rate, temperature and pressure). Selected operating points are the most dominant operating points related to the marine steam turbine operation because steam operating parameters at the turbine inlet and outlet also gives information about steam generators and main steam condenser operation. Additionally, knowing steam operating parameters in only one extraction, mounted between turbine cylinders, will be satisfactory for MLP estimation of turbine exergy analysis

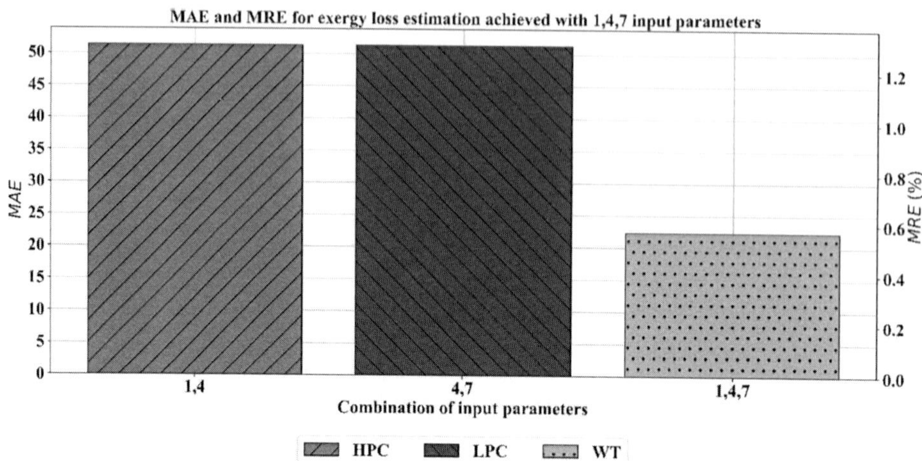

Figure 20. *MAE and MRE values of the selected input combination* $(1, 4, 7)$ *for exergy destruction (exergy loss).*

parameters. Therefore, MLP application can be very beneficial for reducing the costs of measurement and regulation equipment. The same idea, applied to the main marine steam turbine and its cylinders in this paper, can be applied to the whole marine steam propulsion plant.

Figure 20 demonstrates *MAE* and MRE of the selected input combinations. All the selected models achieve the errors of below 1.5%, with the models for HPC and LPC exergy destruction estimation achieving MRE of 1.2%, and the WT exergy destruction model achieving the MRE of 0.6%.

R^2 scores for exergy loss of the whole turbine and each cylinder, presented in Figure 21, show that the models for HPC and WT achieve high fidelity with R^2 scores in excess of 0.99. The lowest R^2 score is achieved by the LPC model which reach the R^2 score of 0.97 . While lower than some scores, when coupled with the low error seen in Figure 20, it can be concluded that this model is satisfactory. The importance of using multiple metrics when evaluating the artificial intelligence-based models is shown here, as models with a high R^2 value may achieve a critically low error or vice-versa. Similarly, a model that may seem to achieve a relatively poor score when a single metric is used may show good performance when evaluated with other metrics, leading to the conclusion that such a model is still usable (providing all the metrics are within the satisfactory ranges).

By observing Figure 22, MAE and MRE values of selected models for exergy efficiency can be seen. As shown, all the models achieve a low error (below 1%), with the error for the HPC exergy efficiency model being less than 0.5% and WT exergy efficiency model showing the error of below 0.3%. The highest error is achieved for the LPC exergy efficiency model, which achieves an error of 0.8%, what is well within the satisfactory error range.

Finally, Figure 23 demonstrates the R^2 scores for the exergy efficiency estimation. With all the scores being in excess of 0.99, it can be concluded that all the

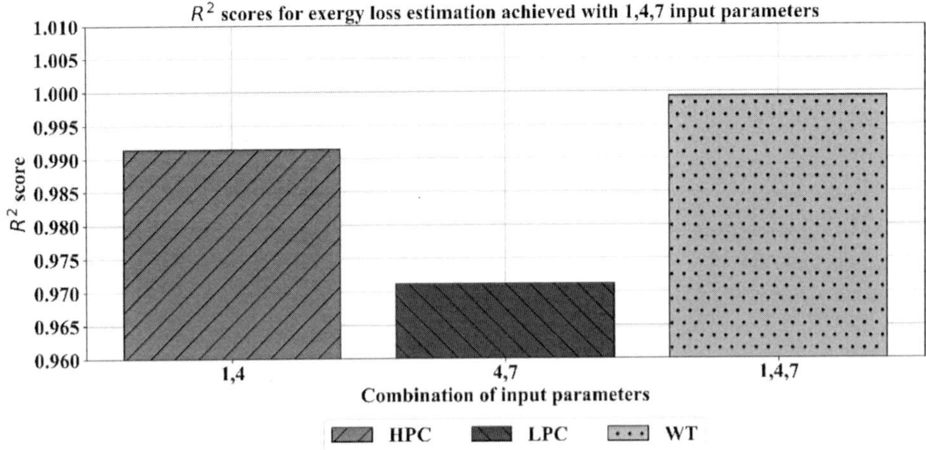

Figure 21. *R^2 values of the selected input combination $(1, 4, 7)$ for exergy destruction (exergy loss).*

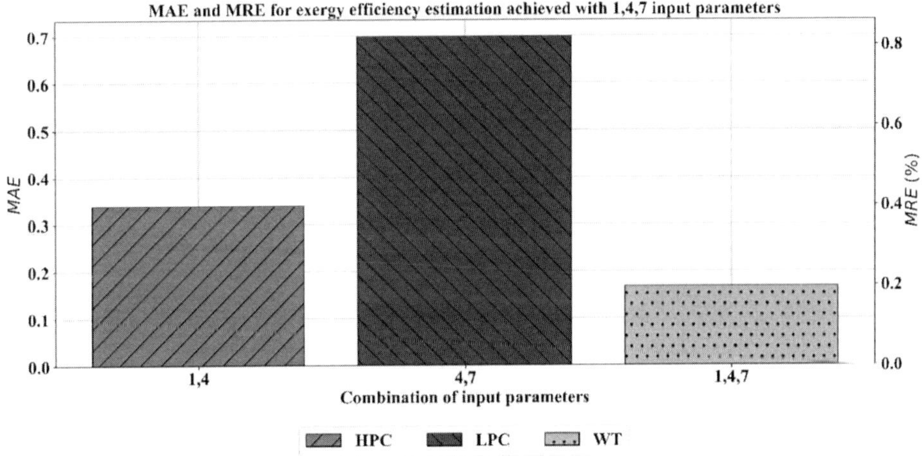

Figure 22. *MAE and MRE values of the selected input combination $(1, 4, 7)$ for exergy efficiency.*

selected models for exergy efficiency estimation of LPC, HPC and WT achieve satisfactory results. All the results are relatively close, with the lowest being the result for the HPC exergy efficiency model, achieving a score just slightly lower than 0.99.

While individually the input parameters may not have the best scores achieved, they all achieve relatively high scores. Considering that errors of models are below 1.5% when these input combinations are used, this is satisfactory for the measurement needs, especially since only three operating points are used. The hyperparameter values and numerical values of metrics for exergy destruction achieved are given in Table 8, and the values for the hyperparameters used in the best

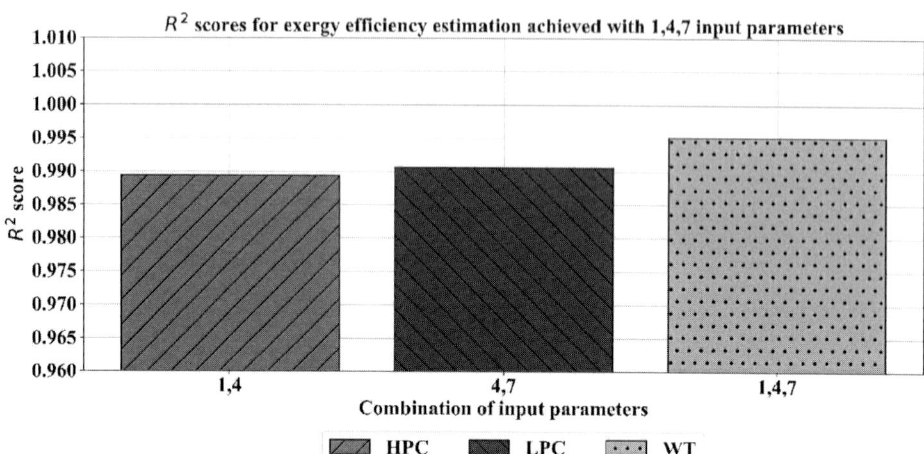

Figure 23. *R^2 values of the selected input combination $(1,4,7)$ for exergy efficiency.*

Table 8. *Metrics achieved by selected operating points, for modeling of exergy destruction.*

Operating Point	R^2	+/-	*MAE*	+/-
1,4 (HPC)	0.9914541639	0.02378781401	51.27713297	36.21271356
4,7 (LPC)	0.9712139368	0.09161658705	36.22307872	50.75836187
1,4,7 (WT)	0.9992643028	0.00172978046	20.44955009	12.32598431

Table 9. *Hyperparameters used for best solutions achieved in selected operating points, for modeling of exergy destruction.*

Operating Point	1,4 (HPC)	4,7 (LPC)	1,4,7 (WT)
Activation Function	ReLU	ReLU	ReLU
L2 Regularization	0.001	0.01	0.1
Hidden Layer Sizes	(84)	(84, 84, 84, 84)	(84, 84, 84, 84)
Learning Rate Type	Adaptive	Constant	Adaptive
Initial learning rate	0.1	0.01	1e-05
Solver	LBFGS	LBFGS	LBFGS

models are presented in Table 9. In the same way, values are given for exergy efficiency in Tables 10 and 11.

Table 8 demonstrates the scores achieved for the exergy destruction with the selected input operating point combinations. It can be seen that the HPC and WT outputs modeled with operating points 1,4 and $1,4,7$, respectively, achieve R^2 scores in excess of 0.99 , while the model for the LPC output, modeled with operating points 4,7 , achieves a lower, but sill satisfactory, R^2 score of 0.97 . Maximal error ranges achieved during the cross-validation have also been provided next to the average of all 10 validation scores for both R^2 and MAE.

Table 10. *Metrics achieved by selected operating points, for modeling of exergy efficiency.*

Operating Point	R^2	+/-	MAE	+/-
1,4 (HPC)	0.9894154031	0.03141286439	0.3393632265	0.3058120431
4,7 (LPC)	0.9906770758	0.02055184947	0.6985228666	1.1429986910
1,4,7 (WT)	0.9951798924	0.01330619104	1.6066829045	0.0133061910

Table 11. *Hyperparameters used for best solutions achieved in selected operating points, for modeling of exergy efficiency.*

Operating Point	1,4 (HPC)	4,7 (LPC)	1,4,7 (WT)
Activation Function	ReLU	ReLU	ReLU
L2 Regularization	0.1	0.1	0.1
Hidden Layer Sizes	(84,42,21)	(84,84,84)	(84,84,84,84)
Learning Rate Type	Constant	Constant	Adaptive
Initial learning rate	0.1	0.01	1e-5
solver	LBFGS	LBFGS	LBFGS

Table 9 provides the hyperparameters used to obtain models that have provided the solutions within the Table 8. The values of hyperparameters presented within Table 9 are a subset of hyperparameters provided within Table 5. Only the hyperparameters which have been used to obtain the best selected models are presented for brevity. It can be seen that a relatively large network, of 4 layers with 84 neurons each, has been used in the case of LPC and WT exergy destruction models, while the HPC exergy destruction model utilized a significantly smaller network-consisting of a single layer with 84 neurons. All the best models for exergy destruction used LBFGS solver and ReLU activation function.

Table 10 presents the scores for the exergy efficiency models, providing R^2 and MAE scores for LPC, HPC and WT models (input combinations being 1,4, 4,7 and 1,4,7 respectively). The table demonstrates that the R^2 scores achieved are in excess of 0.99 for LPC and WT exergy efficiency models, while the HPC model achieved the average R^2 score of 0.989 after the cross-validation process. MAE scores demonstrate an error of less than 1% of exergy efficiency for HPC and LPC models. MAE score grows to 1.6% for WT exergy efficiency, but the error is still within the previously stated satisfactory error range. As was the case in Table 8 , the maximal error ranges obtained during the cross-validation are provided next to the obtained average R^2 and MAE scores in Table 10.

Hyperparameters used to obtain the solutions in the case of exergy efficiency are given in Table 11. As in Table 9, it can be seen that ReLU activation function and LBFGS solver have been used in all cases. Interestingly, the initial learning rates for each individual input operating point combination are the same as they were for exergy destruction models, indicating a similarity between the two regression problems. By observing the hidden layer size hyperparameter, it can be seen that the model for HPC exergy efficiency also required a lower number of neurons in comparison to the LPC and WT exergy efficiency models, as it did in the models of exergy destruction. Still, the number of neurons for HPC exergy efficiency model is much closer to the LPC and WT exergy efficiency models than it was the case for exergy destruction, indicating closer complexity in the problems.

From the hyperparameter tables (Tables 9 and 11), it can be seen that all solutions tended towards the larger number of hidden neurons. This, coupled with a high initial learning rate which was kept either constant or adaptive, signifies a hard to model problem [84]. It can also be seen that some models, namely the ones pertaining to the HPC exergy destruction and exergy efficiency, used smaller networks. The hidden layer sizes were (84) and (84, 42, 21) for exergy destruction and exergy efficiency, compared to the hidden layer sizes for other models, (84, 84, 84) and (84, 84, 84, 84), which are significantly higher. This may indicate that the HPC models were simpler to regress [84,93]. Still, it should be noted that the hidden layer values have tended towards the upper range of the possible hyperparameter values, indicating a relatively complex regression problem. The regularization hyperparameter value is also the highest possible across all models, indicating that some of the inputs used initially had a larger influence on the output, which required lowering in order to achieve a precise model. It should be noted that this may not indicate an operating point by itself as an input, but one of the values (steam temperature, pressure or mass flow rate) that were measured in it. Additionally, it can be seen that across all models, regularization is kept high, the preferred solver is LBFGS and the activation function is ReLU in all cases. While these hyperparameters do not provide additional information about the problem complexity, they can be utilized in further research in order to lower the number of combinations in the GS algorithm and allow for faster training times.

7. CONCLUSIONS

From the presented results it can be concluded that the calculation of exergy destruction and exergy efficiency of a marine steam turbine and its cylinders can be performed using artificial intelligence methods, namely MLP. Through the analysis of the results it can be seen that, while the best results are still achieved when using all the operating points, good results can be achieved using only a subset with negligible loss in accuracy and precision when MLP is used for modeling.

By observing the hyperparameter values which defined the model architecture of the selected models it can be concluded that the best solutions used relatively complex architectures, pointing towards a high complexity of this problem. While it is possible that even more precise solutions could be found using even more complex architectures, as the results provided are within the satisfactory error range, a further increase of computational complexity is needless.

Furthermore, by selecting a subset of the inputs which achieves relatively high results in all six output measurements (exergy destruction and exergy efficiency for LPC, HPC and WT) it can be concluded that, through the application of an MLP model, the number of operating points which can be used to determine the outputs can be lowered. This means that the costs of measuring equipment can be significantly lowered when this method is utilized. This is especially apparent in the selected case of operating points used being 1, 4 and 7 considering that measuring points 1 and 7 are already equipped through the subsequent and prior equipment in the maritime environment that the turbine is placed in. While the selected points do not provide the best results within the tested operating point input combinations, they provide good results across the entire range of modeled outputs and allow for the lowest amount of measuring equipment to be utilized. It may be possible to find an input combination that provides an even better fit for the presented problem. However, it should be noted that this could only be done by performing an extensive search of all possible operating point combinations as inputs, which would significantly raise the computational complexity, when the presented methodology is applied. The increase in computational complexity, in combination with the fact that models obtained by selecting a subset of input combinations provide a high quality solution, means that such an approach is unnecessary for the presented problem.

Further research will be performed in the following direction:

1 To determine optimal turbine operating points for the measurement of steam temperature, pressure and mass flow rate. The goal will be to find three or four operating points of which the measurement results, along with MLP application, can be used for exergy analysis parameters prediction of the whole turbine and each cylinder at any load, with the lowest possible errors.
2 Extensive measurements during a long time period will allow determining performance degradation coefficients for the whole analyzed turbine and each of its cylinders. Implementation of such coefficients inside MPL structure will allow accurate and precise predicting of turbine exergy analysis parameters for the entire period of its operation.
3 Investigate if the same technique can be applied for other main marine steam turbines (especially for newer variants, which consist of three cylinders and steam reheating).

The final goal will be to reduce the number of measurements (and to reduce the number of measuring equipment) for new ships with steam propulsion. The main idea is to perform MLP neural network training and testing on the manufacturer's test data at various loads along with the implementation of performance degradation coefficients inside the MLP structure. Such an approach will allow that onboard the ship, measurements of steam temperature, pressure and mass flow rate can be performed only in three or four optimal operating points (not in seven operating points as at the moment). By using the measurement results from a reduced number of operating points, the MLP neural network will be used for predicting the main steam turbine (and each cylinder) exergy destruction and exergy efficiency. In the end, the application of performance degradation coefficients will allow accurate and precise prediction of the whole turbine and its cylinders exergy analysis parameters for the entire steam power plant operation period.

If possible, the same idea presented in this paper for the main marine steam turbine and its cylinders will be applied to the entire marine steam propulsion plant.

Author Contributions

Conceptualization, S.B.Š., I.L., N.A., V.M. and Z.C.; methodology, V.M., I.L. and S.B.Š.; software, S.B.Š., N.A. and Z.C.; validation, I.L. and V.M.; formal analysis, N.A.; investigation, V.M., I.L., S.B.Š. and N.A.; resources, Z.C.; data curation, V.M.; writing-original draft preparation, V.M., I.L. and S.B.Š.; writing-review and editing, V.M, I.L., N.A. and S.B.Š.; visualization, I.L.; supervision, Z.C. and V.M.; project administration, Z.C.; funding acquisition, V.M. and Z.C. All authors have read and agreed to the published version of the manuscript.

APPENDIX A SPECIFICATION OF USED MEASURING EQUIPMENT

Table A1. *Temperature Measurements.*

→ Greisinger GTF 601-Pt100	
Measuring range:	**−200 to +600 °C**
Response time:	approximate 10 s
Standard:	1/3 DIN class B
Error ranges:	$\pm(0.10 + 0.00167\lvert\text{Temp. in }°\text{C}\rvert)$
→ Greisinger GTF 401-Pt100	
Measuring range:	**−50 to +400 °C**
Response time:	approximate 10 s
Standard:	DIN class B
Error ranges:	$\pm(0.30 + 0.00500\cdot\lvert\text{Temp. in }°\text{C}\rvert)$

Table A2. *Pressure Measurements.*

→ Yamatake JTG960A	
Measuring span:	**0.7 to 14 MPa**
Setting range:	−0.1 to 14 MPa
Working pressure range:	2.0 kPa to 14 MPa
Accuracy:	±0.15% for $\psi \geq 2.1$ MPa $\pm\left(0.05 + 0.1\cdot\frac{2.1}{\psi}\right)$% for $\psi < 2.1$ MPa
→ Yamatake JTG940A	
Measuring span:	**35 to 3500 kPa**
Setting range:	−100 to 3500 kPa
Working pressure range:	2.0 kPa to 3500 kPa
Accuracy:	±0.1% for $\psi \geq 0.14$ MPa $\pm\left(0.025 + 0.75\cdot\frac{0.14}{\psi}\right)$% for $\psi < 0.14$ MPa

ψ = upper and lower limit of the calibration range (for both pressure measuring devices).

Table A3. *Mass Flow Rate Measurements.*

→ Yamatake JTG960A	
Measuring span:	**0.25 to 14 MPa**
Setting span:	−100 to 14 MPa
Working pressure range:	2.0 kPa to 14 MPa
Accuracy:	±0.15% for $\psi \geq 3.5$ MPa $\pm\left(0.1 + 0.05 \cdot \frac{3.5}{\psi}\right)\%$ for $\psi < 3.5$ MPa
→ Yamatake JTD930A	
Measuring span:	**35 to 700 kPa**
Setting span:	−100 to 700 kPa
Working pressure range:	2.0 kPa to 14 MPa
Accuracy:	±0.1% for $\psi \geq 140$ kPa $\pm\left(0.025 + 0.075 \cdot \frac{140}{\psi}\right)\%$ for $\psi < 140$ kPa
→ Yamatake JTD920A	
Measuring span:	**0.75 to 100 kPa**
Setting span:	−100 to 100 kPa
Working pressure range:	2.0 kPa to 14 MPa
Accuracy:	±0.1% for $\psi \geq 5.0$ kPa $\pm\left(0.025 + 0.075 \cdot \frac{5.0}{\psi}\right)\%$ for $\psi < 5.0$ kPa
→ Yamatake JTD910A	
Measuring span:	**0.1 to 2 kPa**
Setting span:	−1 to 1 kPa
Working pressure range:	up to 210 kPa
Accuracy:	$\pm\left(0.15 + 0.15 \cdot \frac{1.0}{\psi}\right)\%$

ψ = upper and lower limit of the calibration range (for all mass flow rate measuring devices).

REFERENCES

1. Tontu, M.; Sahin, B.; Bilgili, M. An exergoeconomic-environmental analysis of an organic Rankine cycle system integrated with a 660MW steam power plant in terms of waste heat power generation. Energy Sources Part A Recovery Util. Environ. Eff. 2020, 2020, 1-22.
2. Elhelw, M.; Al Dahma, K.S. Utilizing exergy analysis in studying the performance of steam power plant at two different operation mode. Appl. Therm. Eng. 2019, 150, 285-293.
3. Uysal, C.; Kurt, H.; Kwak, H.Y. Exergetic and thermoeconomic analyses of a coal-fired power plant. Int. J. Ther. Sci. 2017, 117, 106-120.

4. Naserbegi, A.; Aghaie, M.; Minuchehr, A.; Alahyarizadeh, G. A novel exergy optimization of Bushehr nuclear power plant by gravitational search algorithm (GSA). Energy 2018, 148, 373-385.
5. Wilding, P.R.; Murray, N.R.; Memmott, M.J. The use of multi-objective optimization to improve the design process of nuclear power plant systems. Ann. Nucl. Energy 2020, 137, 107079.
6. Adibhatla, S.; Kaushik, S.C. Exergy and thermoeconomic analyses of 500 MWe sub critical thermal power plant with solar aided feed water heating. Appl. Therm. Eng. 2017, 123,340-352.
7. Mehrpooya, M.; Ghorbani, B.; Sadeghzadeh, M. Hybrid solar parabolic dish power plant and high-temperature phase change material energy storage system. Int. J. Energy Res. 2019, 43, 5405-5420.
8. Idris, M.N.M.; Hashim, H.; Razak, N.H. Spatial optimisation of oil palm biomass co-firing for emissions reduction in coal-fired power plant. J. Clean. Prod. 2018, 172, 3428-3447.
9. Kim, D.; Kim, K.T.; Park, Y.K. A Comparative Study on the Reduction Effect in Greenhouse Gas Emissions between the Combined Heat and Power Plant and Boiler. Sustainability 2020, 12, 5144.
10. Li, X.; Teng, Y.; Zhang, K.; Peng, H.; Cheng, F.; Yoshikawa, K. Mercury Migration Behavior from Flue Gas to Fly Ashes in a Commercial Coal-Fired CFB Power Plant. Energies 2020, 13, 1040.
11. Nazir, S.M.; Bolland, O.; Amini, S. Analysis of combined cycle power plants with chemical looping reforming of natural gas and pre-combustion CO_2 capture. Energies 2018, 11, 147.
12. Javadi, M.A.; Hoseinzadeh, S.; Ghasemiasl, R.; Heyns, P.S.; Chamkha, A.J. Sensitivity analysis of combined cycle parameters on exergy, economic, and environmental of a power plant. J. Ther. Anal. Calor. 2020, 139, 519-525.
13. Rao, A.G.; Van den Oudenalder, F.S.C.; Klein, S.A. Natural gas displacement by wind curtailment utilization in combined-cycle power plants. Energy 2019, 168, 477-491.
14. Kotowicz, J.; Brzęczek, M. Analysis of increasing efficiency of modern combined cycle power plant: A case study. Energy 2018, 153, 90-99.
15. Pattanayak, L.; Sahu, J.N.; Mohanty, P. Combined cycle power plant performance evaluation using exergy and energy analysis. Env. Progr. Sust. Energy 2017, 36, 1180-1186.
16. Okubo, M.; Kuwahara, T. New Technologies for Emission Control in Marine Diesel Engines; Butterworth-Heinemann: Oxford, UK, 2020.
17. Sartomo, A.; Santoso, B.; Muraza, O. Recent progress on mixing technology for water-emulsion fuel: A review. Energy Convers. Manag. 2020, 213, 112817.
18. Senčić, T.; Mrzljak, V.; Blecich, P.; Bonefačić, I. 2D CFD simulation of water injection strategies in a large marine engine. J. Mar. Sci. Eng. 2019, 7, 296.
19. Lamas Galdo, M.I.; Castro-Santos, L.; Rodriguez Vidal, C.G. Numerical analysis of NOx reduction using ammonia injection and comparison with water injection. J. Mar. Sci. Eng. 2020, 8, 109.
20. Fernández, I.A.; Gómez, M.R.; Gómez, J.R.; Insua, Á.B. Review of propulsion systems on LNG carriers. Renew. Sustain. Energy Rev. 2017, 67, 1395-1411.

21. Ammar, N.R. Environmental and cost-effectiveness comparison of dual fuel propulsion options for emissions reduction onboard LNG carriers. Shipbuilding 2019, 70, 61-77.
22. Altosole, M.; Benvenuto, G.; Zaccone, R.; Campora, U. Comparison of Saturated and Superheated Steam Plants for Waste-Heat Recovery of Dual-Fuel Marine Engines. Energies 2020, 13, 985.
23. Altosole, M.; Benvenuto, G.; Campora, U.; Laviola, M.; Trucco, A. Waste heat recovery from marine gas turbines and diesel engines. Energies 2017, 10, 718.
24. Grzesiak, S.; Adamkiewicz, A. Application of Steam Jet Injector for Latent Heat Recovery of Marine steam Turbine Propulsion Plant. New Trend. Prod. Eng. 2018, 1, 235-244.
25. Marques, C.H.; Caprace, J.D.; Belchior, C.R.; Martini, A. An Approach for Predicting the Specific Fuel Consumption of Dual-Fuel Two-Stroke Marine Engines. J. Mar. Sci. Eng. 2019, 7, 20.
26. Mrzljak, V.; Poljak, I.; Mrakovčić, T. Energy and exergy analysis of the turbo-generators and steam turbine for the main feed water pump drive on LNG carrier. Energy Convers. Manag. 2017, 140, 307-323.
27. Behrendt, C.; Stoyanov, R. Operational characteristic of selected marine turbounits powered by steam from auxiliary oil-fired boilers. New Trend. Prod. Eng. 2018, 1, 495-501.
28. Mrzljak, V. Low power steam turbine energy efficiency and losses during the developed power variation. Tech. J. 2018, 12, 174-180.
29. Tanuma, T. Advances in Steam Turbines for Modern Power Plants; Woodhead Publishing: Cambridge, MA, USA, 2017
30. Sun, L.; Hua, Q.; Shen, J.; Xue, Y.; Li, D.; Lee, K.Y. Multi-objective optimization for advanced superheater steam temperature control in a 300MW power plant. Appl. Energy 2017, 208, 592-606.
31. Szargut, J. Exergy Method-Technical and Ecological Applications; WIT Press: Southampton, UK, 2005.
32. Kanoglu, M.; Çengel, Y.A.; Dincer, I. Efficiency Evaluation of Energy Systems; Springer Briefs in Energy; Springer: Berlin/Heidelberg, Germany, 2012.
33. Ahmadi, G.R.; Toghraie, D. Energy and exergy analysis of Montazeri Steam Power Plant in Iran. Renew Sustain. Energy Rev. 2016, 56, 454-463.
34. Si, N.; Zhao, Z.; Su, S.; Han, P.; Sun, Z.; Xu, J.; Cui, X.; Hu, S.; Wang, Y.; Jiang, L.; et al. Exergy analysis of a 1000 MW double reheat ultra-supercritical power plant. Energy Convers. Manag. 2017, 147, 155-165.
35. Ibrahim, T.K.; Basrawi, F.; Awad, O.I.; Abdullah, A.N.; Najafi, G.; Mamat, R.; Hagos, F.Y. Thermal performance of gas turbine power plant based on exergy analysis. Appl. Therm. Eng. 2017, 115, 977-985.
36. Aghbashlo, M.; Tabatabaei, M.; Hosseini, S.S.; Dashti, B.B.; Soufiyan, M.M. Performance assessment of a wind power plant using standard exergy and extended exergy accounting (EEA) approaches. J. Clean. Prod. 2018, 171, 127-136.
37. AlZahrani, A.A.; Dincer, I. Energy and exergy analyses of a parabolic trough solar power plant using carbon dioxide power cycle. Energy Convers. Manag. 2018, 158, 476-488.

38. Abuelnuor, A.A.A.; Saqr, K.M.; Mohieldein, S.A.A.; Dafallah, K.A.; Abdullah, M.M.; Nogoud, Y.A.M. Exergy analysis of Garri "2" 180 MW combined cycle power plant. Renew. Sustain. Energy Rev. 2017, 79, 960-969.
39. Zhao, Z.; Su, S.; Si, N.; Hu, S.; Wang, Y.; Xu, J.; Jiang, L.; Chen, G.; Xiang, J. Exergy analysis of the turbine system in a 1000MW double reheat ultra-supercritical power plant. Energy 2017, 119, 540-548.
40. Medica-Viola, V.; Mrzljak, V.; Anđelić, N.; Jelić, M. Analysis of Low-Power Steam Turbine with One Extraction for Marine Applications. Our Sea 2020, 67, 87-95.
41. Presciutti, A.; Asdrubali, F.; Baldinelli, G.; Rotili, A.; Malavasi, M.; Di Salvia, G. Energy and exergy analysis of glycerol combustion in an innovative flameless power plant. J. Clean. Prod. 2018, 172, 3817-3824.
42. Szablowski, L.; Krawczyk, P.; Badyda, K.; Karellas, S.; Kakaras, E.; Bujalski, W. Energy and exergy analysis of adiabatic compressed air energy storage system. Energy 2017, 138, 12-18.
43. Arshad, A.; Ali, H.M.; Habib, A.; Bashir, M.A.; Jabbal, M.; Yan, Y. Energy and exergy analysis of fuel cells: A review. Therm. Sci. Eng. Progr. 2019, 9, 308-321.
44. Lorencin, I.; Anđelić, N.; Mrzljak, V.; Car, Z. Exergy analysis of marine steam turbine labyrinth (gland) seals. Sci. J. Mar. Res. 2019, 33, 76-83.
45. Kavian, S.; Aghanajafi, C.; Mosleh, H.J.; Nazari, A.; Nazari, A. Exergy, economic and environmental evaluation of an optimized hybrid photovoltaic-geothermal heat pump system. Appl. Energy 2020, 276, 115469.
46. Nami, H.; Anvari-Moghaddam, A. Geothermal driven micro-CCHP for domestic application-Exergy, economic and sustainability analysis. Energy 2020, 207, 118195.
47. Liu, X.; Yang, X.; Yu, M.; Zhang, W.; Wang, Y.; Cui, P.; Zhu, Z.; Ma, Y.; Gao, J. Energy, exergy, economic and environmental (4E) analysis of an integrated process combining CO_2 capture and storage, an organic Rankine cycle and an absorption refrigeration cycle. Energy Convers. Manag. 2020, 210, 112738.
48. Sun, K.; Wu, X.; Xue, J.; Ma, F. Development of a new multi-layer perceptron based soft sensor for SO_2 emissions in power plant. J. Proc. Control 2019, 84, 182-191.
49. Hamed, W.; Salim, N. Use Data Mining Techniques to Identify Parameters That Influence Generated Power in Thermal Power Plant. JECS 2017, 17, 52-64. Available online: http://journal.sustech.edu/index.php/JECS/ article/view/165 (accessed on 7 October 2020).
50. Lorencin, I.; Anđelić, N.; Mrzljak, V.; Car, Z. Genetic algorithm approach to design of multi-layer perceptron for combined cycle power plant electrical power output estimation. Energies 2019, 12, 4352.
51. Khademi, M.; Moadel, M.; Khosravi, A. Power prediction and technoeconomic analysis of a solar PV power plant by MLP-ABC and COMFAR III, considering cloudy weather conditions. Int. J. Chem. Eng. 2016, 2016, 1031943.
52. Demirdelen, T.; Aksu, I.O.; Esenboga, B.; Aygul, K.; Ekinci, F.; Bilgili, M. A New Method for Generating Short-Term Power Forecasting Based on Artificial Neural Networks and Optimization Methods for Solar Photovoltaic Power Plants; Springer Nature Singapore Pte Ltd.: Singapore, 2019.
53. Wahid, F.; Ghazali, R.; Shah, A.S.; Fayaz, M. Prediction of energy consumption in the buildings using multi-layer perceptron and random forest. IJAST 2017, 101, 13-22.

54. Tahan, M.; Tsoutsanis, E.; Muhammad, M.; Karim, Z.A. Performance-based health monitoring, diagnostics and prognostics for condition-based maintenance of gas turbines: A review. Appl. Energy 2017, 198, 122-144.

55. Lorencin, I.; Anđelić, N.; Mrzljak, V.; Car, Z. Multilayer perceptron approach to condition-based maintenance of marine CODLAG propulsion system components. Sci. J. Mar. Res. 2019, 33, 181-190.

56. Ferrero Bermejo, J.; Gómez Fernández, J.F.; Pino, R.; Crespo Márquez, A.; Guillén López, A.J. Review and Comparison of Intelligent Optimization Modelling Techniques for Energy Forecasting and Condition-Based Maintenance in PV Plants. Energies 2019, 12, 4163.

57. Dixit, S.; Verma, N.K. Intelligent Condition Based Monitoring of Rotary Machines with Few Samples. IEEE Sens. J. 2020, 2020.

58. Baressi Šegota, S.; Lorencin, I.; Musulin, J.; Štifanić, D.; Car, Z. Frigate Speed Estimation Using CODLAG Propulsion System Parameters and Multilayer Perceptron. Our Sea 2020, 67, 117-125.

59. Dhini, A.; Kusumoputro, B.; Surjandari, I. Neural network based system for detecting and diagnosing faults in steam turbine of thermal power plant. In Proceedings of the 2017 IEEE 8th International Conference on Awareness Science and Technology (iCAST), Taichung, Taiwan, China, 8-10 November 2017; pp. 149-154.

60. Tian, D.; Deng, J.; Vinod, G.; Santhosh, T.V.; Tawfik, H. A Neural Networks Design Methodology for Detecting Loss of Coolant Accidents in Nuclear Power Plants. In Applications of Big Data Analytics; Springer: Cham, Switzerland, 2018; pp. 43-61.

61. Ayo-Imoru, R.M.; Cilliers, A.C. Continuous machine learning for abnormality identification to aid condition-based maintenance in nuclear power plant. Ann. Nucl. Energy 2018, 118, 61-70.

62. Strušnik, D.; Avsec, J. Artificial neural networking and fuzzy logic exergy controlling model of combined heat and power system in thermal power plant. Energy 2015, 80,318-330.

63. Strušnik, D.; Golob, M.; Avsec, J. Artificial neural networking model for the prediction of high efficiency boiler steam generation and distribution. Simul. Model. Pract. Theory 2015, 57, 58-70.

64. Agrež, M.; Avsec, J.; Strušnik, D. Entropy and exergy analysis of steam passing through an inlet steam turbine control valve assembly using artificial neural networks. Int. J. Heat Mass Transf. 2020, 156, 119897.

65. Marine Steam Turbine MS40-2-Instruction Book for Marine Turbine Unit; Hyundai-Mitsubishi, Hyundai Heavy Industries, Co., Ltd.: Ulsan, Korea, 2004.

66. Mrzljak, V.; Poljak, I.; Medica-Viola, V. Dual fuel consumption and efficiency of marine steam generators for the propulsion of LNG carrier. Appl. Therm. Eng. 2017, 119, 331-346.

67. Koroglu, T.; Sogut, O.S. Conventional and advanced exergy analyses of a marine steam power plant. Energy 2018, 163, 392-403.

68. Çiçek, A.N. Exergy Analysis of a Crude Oil Carrier Steam Plant. Master's Thesis, Istanbul Technical University, Istanbul, Turkey, 2009. (In Turkish)

69. Mrzljak, V.; Poljak, I.; Medica-Viola, V. Thermodynamical analysis of high-pressure feed water heater in steam propulsion system during exploitation. Shipbuilding 2017, 68, 45-61.

70. Taylor, D.A. Introduction to Marine Engineering, 2nd ed.; Elsevier Butterworth-Heinemann: Oxford, UK, 1996.
71. Škopac, L.; Medica-Viola, V.; Mrzljak, V. Selection Maps of Explicit Colebrook Approximations according to Calculation Time and Precision. Heat Transf. Eng. 2020, 2020, 1-15.
72. Carlton, J. Marine Propellers and Propulsion, 4th ed.; Butterworth-Heinemann: Oxford, UK, 2019.
73. Kocijel, L.; Poljak, I.; Mrzljak, V.; Car, Z. Energy Loss Analysis at the Gland Seals of a Marine Turbo-Generator Steam Turbine. Tech. J. 2020, 14, 19-26.
74. Moran, M.; Shapiro, H.; Boettner, D.D.; Bailey, M.B. Fundamentals of Engineering Thermodynamics, 7th ed.; John Wiley and Sons, Inc.: Hoboken, NJ, USA, 2011.
75. Fernández, I.A.; Gómez, M.R.; Gómez, J.R.; López-González, L.M. H_2 production by the steam reforming of excess boil off gas on LNG vessels. Energy Convers Manag. 2017, 134, 301-313.
76. Medica-Viola, V.; Baressi Šegota, S.; Mrzljak, V.; Štifanić, D. Comparison of conventional and heat balance based energy analyses of steam turbine. Sci. J. Mar. Res. 2020, 34, 74-85.
77. Dincer, I.; Rosen, M.A. Exergy: Energy, Environment and Sustainable Development, 2nd ed.; Elsevier: Oxford, UK, 2013.
78. Baldi, F.; Ahlgren, F.; Nguyen, T.V.; Thern, M.; Andersson, K. Energy and exergy analysis of a cruise ship. Energies 2018, 11, 2508.
79. Kumar, V.; Pandya, B.; Matawala, V. Thermodynamic studies and parametric effects on exergetic performance of a steam power plant. Int. J. Ambient. Energy 2019, 40, 1-11.
80. Mrzljak, V.; Blecich, P.; Anđelić, N.; Lorencin, I. Energy and exergy analyses of forced draft fan for marine steam propulsion system during load change. J. Mar. Sci. Eng. 2019, 7, 381.
81. Ray, T.K.; Datta, A.; Gupta, A.; Ganguly, R. Exergy-based performance analysis for proper O&M decisions in a steam power plant. Energy Convers. Manag. 2010, 51, 1333-1344.
82. Aljundi, I.H. Energy and exergy analysis of a steam power plant in Jordan. Appl. Therm. Eng. 2009, 29, 324-328.
83. Mrzljak, V.; Senčić, T.; Žarković, B. Turbogenerator Steam Turbine Variation in Developed Power: Analysis of Exergy Efficiency and Exergy Destruction Change. Model. Simul. Eng. 2018, 2018, 2945325.
84. Tan, H.; Shan, S.; Nie, Y.; Zhao, Q. A new boil-off gas re-liquefaction system for LNG carriers based on dual mixed refrigerant cycle. Cryogenics 2018, 92, 84-92.
85. Noroozian, A.; Mohammadi, A.; Bidi, M.; Ahmadi, M.H. Energy, exergy and economic analyses of a novel system to recover waste heat and water in steam power plants. Energy Convers. Manag. 2017, 144, 351-360.
86. Nanaki, E.A.; Xydis, G. Exergetic Aspects of Renewable Energy Systems: Insights to Transportation and Energy Sector for Intelligent Communities; CRC Press: Boca Raton, FL, USA, 2019.
87. Erdem, H.H.; Akkaya, A.V.; Cetin, B.; Dagdas, A.; Sevilgen, S.H.; Sahin, B.; Teke, I.; Gungor, C.; Atas, S. Comparative energetic and exergetic performance analyses for coal-fired thermal power plants in Turkey. Int. J. Therm. Sci. 2009, 48, 2179-2186.

88. Adibhatla, S.; Kaushik, S.C. Energy and exergy analysis of a super critical thermal power plant at various load conditions under constant and pure sliding pressure operation. Appl. Therm. Eng. 2014, 73, 51-65.

89. Goodfellow, I.; Bengio, Y.; Courville, A. Deep Learning; The MIT Press: Cambridge, MA, USA, 2016.

90. Moon, T.; Hong, S.; Choi, H.Y.; Jung, D.H.; Chang, S.H.; Son, J.E. Interpolation of greenhouse environment data using multilayer perceptron. Comput. Electron. Agric. 2019, 166, 105023.

91. Lorencin, I.; Anđelić, N.; Španjol, J.; Car, Z. Using multi-layer perceptron with Laplacian edge detector for bladder cancer diagnosis. Artif. Intell. Med. 2020, 102, 101746.

92. Car, Z.; Baressi Šegota, S.; Anđelić, N.; Lorencin, I.; Mrzljak, V. Modeling the Spread of COVID-19 Infection Using a Multilayer Perceptron. Comput. Math. Methods Med. 2020, 2020, 5714714.

93. Khalid, A.; Sundararajan, A.; Acharya, I.; Sarwat, A.I. Prediction of li-ion battery state of charge using multilayer perceptron and long short-term memory models. In Proceedings of the 2019 IEEE Transportation Electrification Conference and Expo (ITEC), Detroit, MI, USA, 19-21 June 2019; pp. 1-6.

94. Bisong, E. The Multilayer Perceptron (MLP). Building Machine Learning and Deep Learning Models on Google Cloud Platform; Apress: Berkeley, CA, USA, 2019; pp. 401-405.

95. Eger, S.; Youssef, P.; Gurevych, I. Is it time to swish? Comparing deep learning activation functions across NLP tasks. arXiv 2019, arXiv:1901.02671.

96. Jagtap, A.D.; Kawaguchi, K.; Karniadakis, G.E. Locally adaptive activation functions with slope recovery term for deep and physics-informed neural networks. Proc. R. Soc. A 2020, 476, 20200334.

97. Dureja, A.; Pahwa, P. Analysis of non-linear activation functions for classification tasks using convolutional neural networks. Recent Pat. Comput. Sci. 2019, 12, 156-161.

98. Hastie, T.; Tibshirani, R.; Friedman, J. The Elements of Statistical Learning: Data Mining, Inference, and Prediction, 2nd ed.; Springer Science & Business Media: New York, NY, USA, 2009.

99. Pedregosa, F.; Varoquaux, G.; Gramfort, A.; Michel, V.; Thirion, B.; Grisel, O.; Blondel, M.; Prettenhofer, P.; Weiss, R.; Dubourg, V.; et al. Scikit-learn: Machine learning in Python. J. Mach. Learn. Res. 2011, 12, 2825 – 2830

100. Abraham, A.; Pedregosa, F.; Eickenberg, M.; Gervais, P.; Mueller, A.; Kossaifi, J.; Gramfort, A.; Thirion, B.; Varoquaux, G. Machine learning for neuroimaging with scikit-learn. Front. Neuroinf 2014, 8, 14.

101. Géron, A. Hands-On Machine Learning with Scikit-Learn, Keras, and TensorFlow: Concepts, Tools, and Techniques to Build Intelligent Systems, 2nd ed.; O'Reilly Media: Sebastopol, CA, USA, 2019.

102. Liashchynskyi, P.; Liashchynskyi, P. Grid Search, Random Search, Genetic Algorithm: A Big Comparison for NAS. arXiv 2019, arXiv:1912.06059.

103. Bari, A.H.; Gavrilova, M.L. Multi-layer perceptron architecture for kinect-based gait recognition. In Proceedings of the Computer Graphics International Conference, Calgary, AB, Canada, 17-20 June 2019; Springer: Cham, Siwtzerland; pp. 356-363.

104. Sakar, C.O.; Polat, S.O.; Katircioglu, M.; Kastro, Y. Real-time prediction of online shoppers' purchasing intention using multilayer perceptron and LSTM recurrent neural networks. Neural Comput. Appl. 2019, 31, 6893-6908.

105. Nagelkerke, N.J. A note on a general definition of the coefficient of determination. Biometrika 1991, 78, 691-692.

106. Nakagawa, S.; Johnson, P.C.; Schielzeth, H. The coefficient of determination R^2 and intra-class correlation coefficient from generalized linear mixed-effects models revisited and expanded. J. R. Soc. Interface 2017, 14, 20170213.

107. Qi, J.; Du, J.; Siniscalchi, S.M.; Ma, X.; Lee, C.H. On mean absolute error for deep neural network based vector-to-vector regression. IEEE Signal Proc. Lett. 2020, 27, 1485-1489.

108. Štifanić, D.; Musulin, J.; Miočević, A.; Baressi Šegota, S.; Šubić, R.; Car, Z. Impact of COVID-19 on Forecasting Stock Prices: An Integration of Stationary Wavelet Transform and Bidirectional Long Short-Term Memory. Complexity 2020, 2020, 1846926.

109. Berrar, D. Cross-validation. Encycl. Bioinform. Comput. Biol. 2019, 1, 542-545.

110. Bishop, C.M. Pattern Recognition and Machine Learning; Springer: New York, NY, USA, 2006.

111. Moayedi, H.; Osouli, A.; Nguyen, H.; Rashid, A.S.A. A novel Harris hawks' optimization and k-fold cross-validation predicting slope stability. Eng. Comput. 2019, 2019, 1-11.

112. BURA Supercomputer, Computing Resources. Available online: https:/cnrm.uniri.hr/bura/ (accessed on 30 October 2020).

113. Anaconda Software Distribution. Anaconda Documentation. Anaconda Inc. 2020. Available online: https://docs.anaconda.com/ (accessed on 30 October 2020).

114. Lemmon, E.W.; Huber, M.L.; McLinden, M.O. Reference Fluid Thermodynamic and Transport Properties-REFPROP; Version 9.0, User's Guide; NIST: Gaithersburg, MD, USA, 2010.

115. Mrzljak, V.; Poljak, I.; Žarković, B. Exergy analysis of steam pressure reduction valve in marine propulsion plant on conventional LNG carrier. Our Sea 2018, 65, 24-31.

116. SUITABLE PT100 MEASURING PROBE (4-WIRE). Available online: https://www.greisinger.de/files/upload/en/produkte/kat/k16_011_EN_oP.pdf (accessed on 3 October 2020).

117. JTG Series of Pressure Transmitters. Available online: http://smte.kr/product/data/pdf/pdf_100812100836_ 552363.pdf (accessed on 4 October 2020).

118. JTD Series of Differential Pressure Transmitters. Available online: http://www.krt product.com/krt_Picture/sample/1_spare%20part/yamatake/Fi_ss01/SS2-DST100-0100.pdf (accessed on 3 October 2020).

CHAPTER 2

Multi-Domain Modeling and Analysis of Marine Steam Power System Based on Digital Twin

Guoqing Zeng, Jintao Wang, Lei Zhang, Xuyang Xie, Xuefeng Wang and Guobing Chen

College of Power Engineering, Naval University of Engineering, Wuhan 430033, China

ABSTRACT

The marine steam power system includes a large amount of thermal equipment; meanwhile, the marine environment is harsh and the working conditions change frequently. Operation management involves many disciplines, such as heat, machinery, control, electricity, etc. It is a complex multi-discipline physical system with typical nonlinear, multi-parameter, strong coupling characteristics. In order to realize the health management of a marine steam power system, based on digital twin technology combined with the Modelica language, modular modeling, etc., this paper conducts in-depth research on the multi-domain modeling of the marine steam power system, characteristic analysis of variable working conditions, fault simulation, etc. The analysis results show that the dynamic response trend of the model is consistent with the actual operation, the error of the main steam flow at 1800 s *is the largest and is* −4.9%*, and the error of the main steam flow, steam turbine output power, cooling water outlet temperature and other key parameters is within* ±5%*. Virtual reality mapping between the digital model and the physical equipment is realized, which lays a foundation for mastering the dynamic characteristics of the marine steam power system.*

Keywords: *steam power system; multi-domain modeling; Modelica; dynamic characteristics; digital twin*

1. INTRODUCTION

As a typical, complex and multi-discipline physical system, the marine steam power system includes a large amount of thermal equipment [1], including boilers, main steam turbines, steam turbine generators, condensers, deaerators, feed pumps, etc., which involve electrical, mechanical, control and other fields, and it has typical nonlinear, multi-parameter, strong coupling characteristics. For such a complex multi-discipline physical system, it is difficult to accurately grasp the dynamic characteristics. Different from the thermal equipment of nuclear power plants, the marine steam power equipment needs to change its speed frequently due to the impact of the mission profile during operation; meanwhile, the operating environment is harsh due to the influence of high salt fog, strong wind and waves in marine conditions. Marine equipment frequently need to operate under variable working conditions and in adverse conditions, which increases the load of the power system and also raises the requirements for the operation management of the marine power system. Therefore, it is imperative to carry out an accurate and effective dynamic characteristics analysis, online monitoring and a status evaluation of the marine steam power system under all working conditions.

At present, many scholars have carried out relevant research on marine power system. Nirbito et al. detailed the performance analysis of a propulsion system engine of an LNG tanker using a combined cycle whose components are a gas turbine, a steam turbine, and a heat recovery steam generator. The maximum power, fuel consumption rate and speed of the LNG carrier using a combined cycle under full load conditions are obtained through analysis of the model [2]. Perabo et al. used the Open Simulation Platform (OSP) to carry out digital twin modeling and co-simulation of the power stage, generator set, local controller, etc. of a ship's power propulsion system according to the complex and coupled characteristics of the ship's power system, and analyzed the stability and dynamic characteristics of the system [3]. Mrzljak et al. studied the steam turbine generator and feed water pump turbine of an LNG ship, analyzed their efficiency by measuring the actual operating parameters, and concluded that the energy conversion efficiency of the steam turbine generator could be improved by $1-3\%$ by replacing the steam turbine with a motor to provide power for the feed water pump [4]. Dulau et al. used Matlab/Simulink to simulate high, medium and low pressure steam turbines, analyzed the influence of the reheater, valve opening change and position height change on the system, and proposed that the key to control is to establish a steady-state response when the turbine is connected to the generator and other controllers in a closed loop [5].

In summary, many scholars have carried out systematic research on marine power systems, including modeling methods, mathematical models, engineering applications, etc., but most scholars still carry out single-domain simulation and dynamic characteristic analysis for the subsystems and key equipment in the power system, while there is relatively little research on the coordination, multi-domain coupling, dynamic response and fault simulation of each piece of thermal equipment in the marine power system.

Due to the fact that traditional modeling methods focus on static analysis, it is difficult to analyze dynamic characteristics under variable conditions and fault conditions. Therefore, the establishment of a high-precision digital model of the system is the key to further promoting the digital and intelligent development of marine systems. As a new technology, digital twin is a key technology for realizing the mapping of physical system to a digital model [6]. It makes full use of the physical model, equipment status, operating environment and other data, integrates multi-disciplinary and multi-timescale simulation processes, and completes virtual reality mapping in the information space, thus reflecting the full life cycle of the corresponding physical equipment [7]. Therefore, the combination of multi-domain modeling, modular modeling and other methods to establish a high-precision digital model under the digital twin technology and carry out a dynamic characteristics simulation analysis can not only overcome the shortcomings of traditional modeling methods, but also lay a solid foundation for the successful application of digital twin technology in the marine steam power system.

The combination of multi-disciplinary modeling methods and digital twin technology can give full play to their advantages in modeling and simulation. At present, several scholars have carried out relevant research. Ryan put forward a model-driven method for the online monitoring and predictive maintenance of automobile braking system, established digital twin models of its key components using Ansys and the Modelica language, analyzed the signal response of the braking system under gear/mechanical failure, and laid a foundation for the subsequent diagnosis and prediction of the braking system [8]. Fotias developed the model library of an electrical power system based on the Modelica language, which includes a multi-fidelity, multi-scale battery, power electronics and an electrical propulsion system sub-library, and used this library to build the digital twin model of an electrical power system in the process of product development, aiming to reduce development costs and carry out status monitoring [9]. Zhou pointed out that the satellite communication station is a typical multi-disciplinary physical system, which is composed of six systems, including the antenna system, receiving system, transmitting system, communication system, control system and power supply system, through the demand analysis of the digital twin construction of satellite earth stations; the method of building a digital twin system of satellite communication equipment using MWorks multidomain system modeling platform is proposed. Finally, the effectiveness and feasibility of the digital twin modeling method are verified [10]. Vering et al. applied digital twin technology to the energy recovery devices in air conditioning systems. Based on the Modelica language, they developed their digital twin model to predict the behavior of the physical system. They communicated with the physical model through the Message Queuing Telemetry Transport (MQTT) interface to achieve the virtual reality mapping of the physical model and the digital twin model, which can improve the efficiency of energy recovery devices [11]. In order to monitor the operation status of the space station in real time, Xing Tao et al. established seven subsystem models of dynamics and control, energy, environmental and thermal control, propulsion, information, data management, and measurement and control of the core cabin, test cabin I and test cabin II of the space station based on the Modelica language and the MWorks system simulation platform, covering the general system, subsystems, and key stand-alone equipment, and conducted simulation analyses of the typical working conditions of subsystems, respectively. The functional mock up interface (FMI) standard was adopted to integrate the digital twin models of a single cabin, two cabin, one font and three cabin T-shaped system of a space station, analyze and verify the rendezvous and docking, transposition and other scenarios, and achieve the analysis and verification of the system level, all boundaries and all working conditions of the space station [12].

In a word, the thermal equipment of the marine steam power system is a typical and strong coupling complex system with complex composition and working conditions, and, with the traditional modeling method, it is difficult to satisfy the development requirements of digital and intelligent thermal equipment. In view of the advantages of multi-disciplinary modeling in the modeling of complex physical systems and the wide application of digital twin technology in equipment health management, this paper combines the ideas of multidisciplinary modeling and modular modeling under the digital twin technology system to build a digital model of the main equipment and subsystems of the marine steam power system, and uses the Modelica language and the MWorks system simulation platform to build the digital model of the marine steam power system. It also carries out an accurate and effective dynamic characteristic analysis and fault simulation, which makes a preliminary exploration for the application of digital twin technology in marine steam power systems, and provides support for the health management of marine thermal equipment.

2. MATHEMATICAL MODEL OF MARINE STEAM POWER SYSTEM EQUIPMENT

In order to accurately grasp the dynamic characteristics of the marine steam power system, the model should satisfy the requirements of real-time and accuracy. Therefore, when establishing the mathematical model of the thermal system, we should fully analyze the working principles of the steam turbine, condenser, deaerator, etc. The working principles of the marine steam power system are shown in Figure 1.

2.1. Mathematical Model of Steam Turbine

In order to carry out an accurate thermodynamic analysis of the marine steam turbine, the establishment of its detailed mathematical model is a necessary premise. The thermodynamic analysis of the steam turbine mainly requires analyzing and calculating the steam turbine's operation under rated and variable conditions and determining its inlet and outlet flow and output power according to the design conditions [13].

(1) Mathematical equation of steam pressure and flow:

$$q_a = q_r \beta_p \frac{p_a / p_r}{\sqrt{T_a / T_r}} f(k) \tag{1}$$

$$\beta_1 = \begin{cases} 1 : (p_g / p_a \leq \varepsilon_{cr}) \\ \sqrt{1 - [(\varepsilon_n - \varepsilon_{cr}) \,/\, (1 - \varepsilon_{cr})]^2} : (p_g / p_a > \varepsilon_{cr}) \end{cases} \tag{2}$$

where:

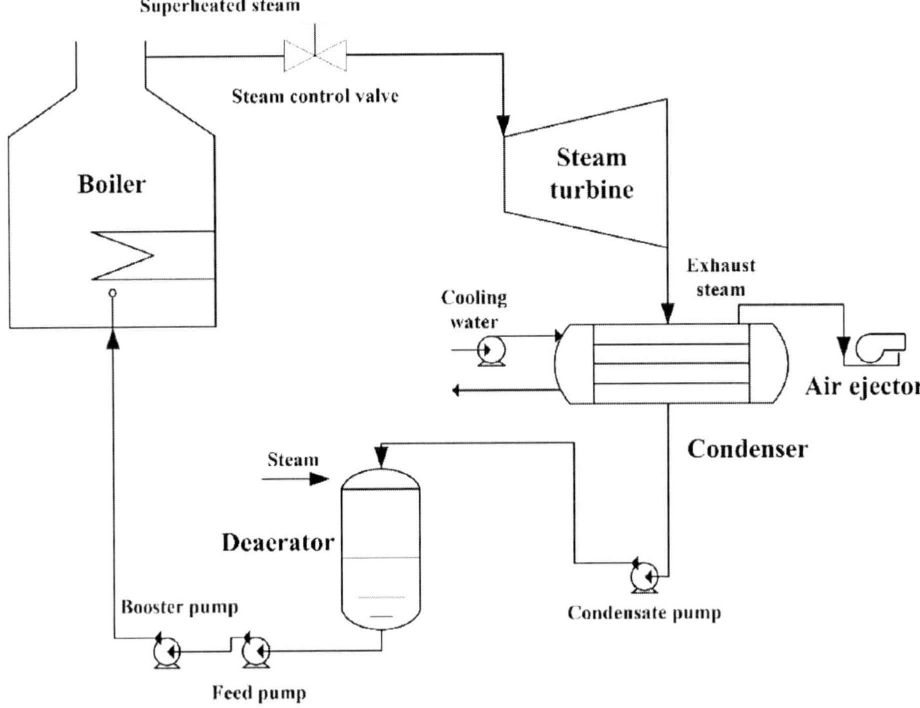

Figure 1. *The working principles of the marine steam power system.*

q_a, q_r-The actual intake of steam, the rated intake of steam, kg/s
p_a, p_r-The actual pressure of steam, the rated pressure of steam, MPa;
T_a, T_r-The actual temperature of steam, the rated temperature of steam, K;
p_g-The outlet pressure of regulating stage, MPa;
β_p-The Pengtemen coefficient;
$f(k)$-The valve characteristic curve function;
ε_{cr}-The critical pressure ratio of governing stage.
(2) Rotor dynamic equation

$$J\frac{d\omega}{dt} = \tau_{in} - \tau_{out} \tag{3}$$

where:
ω-The speed of rotor, rad/s;
J-Moment of inertia, kg $\cdot$ m^2;
τ_{in} -Input torque, kg $\cdot$ m;
τ_{in} -Output torque, kg $\cdot$ m.

2.2. Mathematical Model of Condenser

In order to establish the mathematical model of the condenser, the condenser is divided into a steam area and a cooling water area [14].

1. Steam area:

 (1) Mass conservation equation:

$$\frac{dM_s}{dt} = \frac{d(\rho_s V_s)}{dt} = q_s - q_{air} - q_c \tag{4}$$

where:

M_s - The mass of steam, kg;
ρ_s - The density of steam, kg/m^3; V_s-The volume of steam, m^3;
q_s - The amount of steam discharged into the condenser, kg/s;
q_{air} - The amount of air discharged into the condenser, kg/s;
q_c-The amount of condensate outlet flow, kg/s.
(2) Energy conservation equation:

$$\frac{d(\rho_s V_s h_s)}{dt} = q_s h_s - q_{air} h_{air} - q_c h_c - Q_c \tag{5}$$

where:

h_s-The enthalpy of steam, kJ/kg; h_{air} -The enthalpy extracted by steam jet ejector, kJ/kg;
h_c - The enthalpy of condensate, kJ/kg;
Q_c - The condensation of steam releases heat, kJ/s.

$$Q_c = \alpha_c A_t (T_s - T_t) \tag{6}$$

where:

α_c-The condensation heat release coefficient between steam and cooling tube, kW/ $(\text{m}^2 \cdot {}^\circ\text{C})$; A_t-The external wall area of cooling tube, m^2;
T_s - The temperature of steam, K;
T_t-The temperature of cooling tube, K.

2. Cooling water area:

 (1) Convective heat transfer of cooling water:

$$Q_w = \alpha_w A_s (T_t - T_w) \tag{7}$$

where:

Q_w - The convective heat transfer of cooling water, kJ/s;
α_w-The convective heat transfer coefficient between cooling water and cooling tube, kW/ $(\text{m}^2 \cdot {}^\circ\text{C})$;
A_s-The heat exchange area of cooling pipe, m^2;
T_w - The temperature of cooling water, K.
(2) Energy conservation equation of cooling tube wall:

$$M_t c_t \frac{dT_t}{dt} = Q_c - Q_w \tag{8}$$

where:

M_t-The quality of cooling tube, kg;
c_t-The specific heat capacity of cooling tube, kJ/ (kg · °C).
(3) Mass conservation equation of cooling tube:

$$M_w c_w \frac{dT_w}{dt} = Q_w - q_w c_w \left(T_{f2} - T_{f1}\right) \tag{9}$$

where:
M_w-The quality of cooling water in cooling tube, kg;
c_w-The specific heat capacity of cooling water, kJ/ (kg · °C);
q_w-The flow of cooling water, kg/s;
T_{f2}-The outlet temperature of cooling water, K;
T_{f1}-The inlet temperature of cooling water, K.
(4) The overall heat transfer coefficient:

$$K = K_0 \beta_t \beta_m \phi_c \tag{10}$$

where:
K-The overall heat transfer coefficient of condenser, kW/ $(m^2 \cdot °C)$;
K_0 - The basic heat transfer coefficient, kW/ $(m^2.°C)$;
β_t-The correction coefficient of cooling water inlet temperature;
β_m-The correction coefficient of cooling tube material and wall thickness;
ϕ_c-The cleanliness coefficient of cooling tube. (5) Logarithmic mean temperature difference:

$$\Delta T_{\ln} = \frac{T_{f2} - T_{f1}}{\ln \frac{(T_s - T_{f1})}{(T_s - T_{f2})}} = \frac{\Delta T}{\ln \frac{\Delta T + \delta_t}{\delta_t}} \tag{11}$$

where:
ΔT - The rise in temperature of cooling water, $\Delta T - T_{f2} - T_{f1}$, K;
δ_t-The heat transfer end difference of condenser, K.
(6) Heat transfer end difference:

$$\delta_t = \frac{\Delta T}{e^{\frac{1.18s}{4.16 \times q_{hs}}} - 1} \tag{12}$$

2.3. Mathematical Model of Deaerator

According to the structure composition and working principle of the deaerator, in order to establish the mathematical model of the deaerator, the deaerator is divided into a deaerator area and a water tank area, and its mathematical model is established [15].

(1) Mass conservation equation:

$$\frac{d(\rho_d V_d)}{dt} = q_t + q_o + q_i - q_e \tag{13}$$

where:

V_d - The volume of water in deaerator, m^3;
ρ_d - The density of water in deaerator, kg/m^3;
q_t - The mass flow of steam entering deaerator, kg/s;
q_0-The mass flow of condensate entering deaerator, kg/s;
q_i-The mass flow of water entering deaerator, kg/s;
q_e - The mass flow of outlet feed water, kg/s.
(2) Energy conservation equation:

$$\frac{d\left(\rho_d V_d h_d\right)}{dt} = q_t h_t + q_o h_o + q_a h_a - q_e h_e \tag{14}$$

where:
h_d-The enthalpy of water in deaerator, kJ/kg;
h_t - The enthalpy of steam entering deaerator, kJ/kg;
h_0-The enthalpy of condensate entering deaerator, kJ/kg;
h_i - The enthalpy of feed water entering deaerator, kJ/kg;
h_e - The enthalpy of outlet feed water, kJ/kg.

2.4. Mathematical Model of Pipeline

The pipeline model is mainly used to simulate the flow process of the medium in the system, calculate the fluid flow state parameters in the pipeline equipment, and consider the effects of heat transfer, friction, etc., which can invoke the physical parameter model library for associated use with the medium.

(1) Mass conservation equation:

$$V_p \frac{d\rho_p}{dt} = q_{out} - q_{in} \tag{15}$$

where:
V_p-The volume of pipeline, m^3;
ρ_p-The fluid density of pipeline, kg/m^3; qout-The fluid outlet mass flow, kg/s;
q_{in} -The fluid inlet mass flow, kg/s.
(2) Energy conservation equation:

$$Q_p = q_{out} h_{out} - q_{in} h_{\text{in}} = \alpha_p A_p \left(T_p - T_e\right) \tag{16}$$

where:
Q_p-The heat of pipeline fluid, kJ/s;
h_{out} -The enthalpy of outlet fluid, kJ/kg;
h_{in} -The enthalpy of inlet fluid, kJ/kg;
α_p-The convective heat transfer coefficient between fluid and pipe, $kW/\left(m^2 \cdot {}^\circ C\right)$
A_p - The heat exchange area of pipeline, m^2;
T_p-The temperature of pipeline fluid, K;
T_e-The temperature of pipeline wall, K.
(3) Frictional loss:

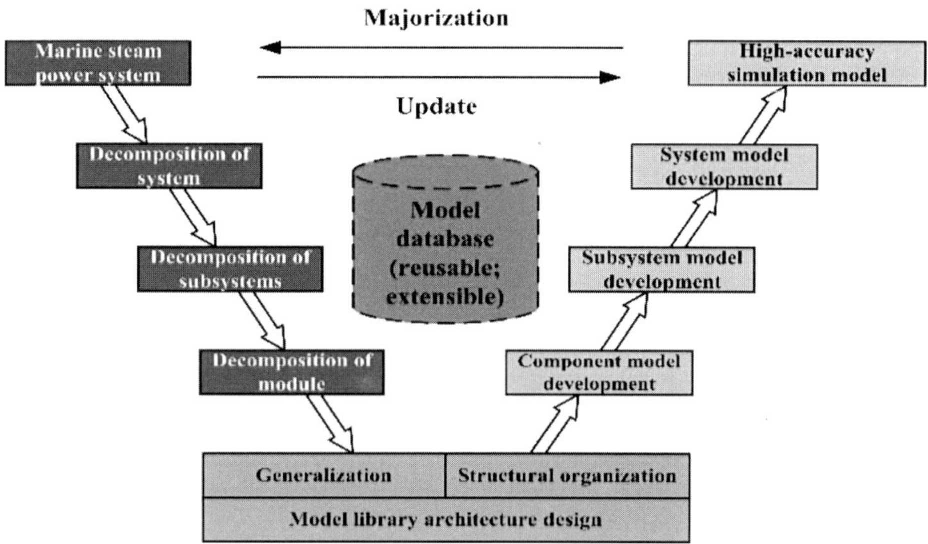

Figure 2. *The model development principle.*

$$\Delta p_{\zeta} = \zeta \frac{l_p}{d_p} \rho_p \frac{v_p^2}{2} \tag{17}$$

where:

ζ-The frictional loss coefficient of pipeline fluid;
l_p-The length of pipeline, m;
d_p-The diameter of pipeline, m;
v_p-The speed of pipeline, m/s;
μ_p-The viscosity coefficient of pipeline fluid;
ε_p-The roughness of pipeline wall.

3. DIGITAL MODELING OF MARINE STEAM POWER SYSTEM BASED ON MODELICA

For each piece of thermal equipment involved in the marine steam power system, including the boilers, main steam turbines, condensers, deaerators, air ejectors, valves, pipes, pumps, etc., according to its mathematical operating characteristics combined with the principle of modular modeling, the corresponding component model and system model library are developed using the MWorks system simulation platform based on the Modelica language, and the digital model of the marine steam power system is integrated. The model development principle is shown in Figure 2.

After the corresponding component models have been developed on the MWorks simulation platform, in order to ensure the accuracy of the established digital model and the high stability of the digital simulation model of the complete steam power system, and to ensure that the model operating state parameters

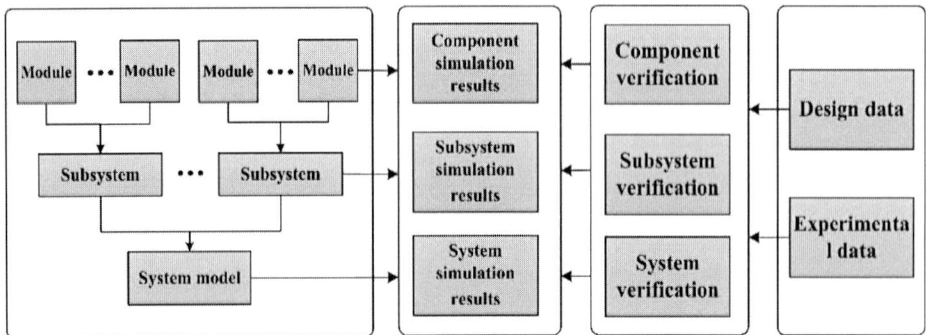

Figure 3. *The model characteristic parameter debugging process.*

can accurately reflect the actual marine operating data, it is necessary to conduct a parameter sensitivity analysis, model modification, a confidence evaluation, etc. on the model. The test data and historical operation data of real marine systems are used to verify and check the model.

The closer the operation data of the digital model are to the actual marine operation data, the higher the accuracy of the model, which can ensure the accuracy of the subsequent dynamic characteristics analysis, realize the online monitoring and status evaluation of the physical model by the digital model, so as to accurately provide an operation strategy for the actual marine thermal equipment, and realize the virtual-real mapping of the digital model to the physical equipment. The debugging process for the model characteristic parameters is shown in Figure 3.

At the same time, in debugging the parameters of the system model, we should focus on whether the data interaction of each component model and subsystem model is normal, whether the model connection mechanism and interface are correct, whether the parameter setting for the model characteristic is reasonable, and whether a dynamic characteristic analysis of the system by the digital model can be realized. Through the debugging of the characteristic parameters of the system model, a complete digital model of the marine steam power system is built. The digital model of the marine steam power system is shown in Figure 4.

Through debugging the design parameters and state parameters in the characteristic parameters of the digital model of the marine steam power system, a more accurate digital model has been obtained. In order to verify the accuracy of the digital model of the marine steam power system, this paper uses a type of steam turbine unit to carry out load up test verification. By controlling the opening of the steam valve, it simulated the processes of acceleration and analyzed their dynamic characteristics. An operating state parameter curve for the marine steam turbine from start-up to a 100% working condition is obtained. All state parameters are dimensionless according to their design data. The dimensionless formula is shown in Formula (18).

$$\widetilde{y}_i = \frac{y_s}{y_d} \tag{18}$$

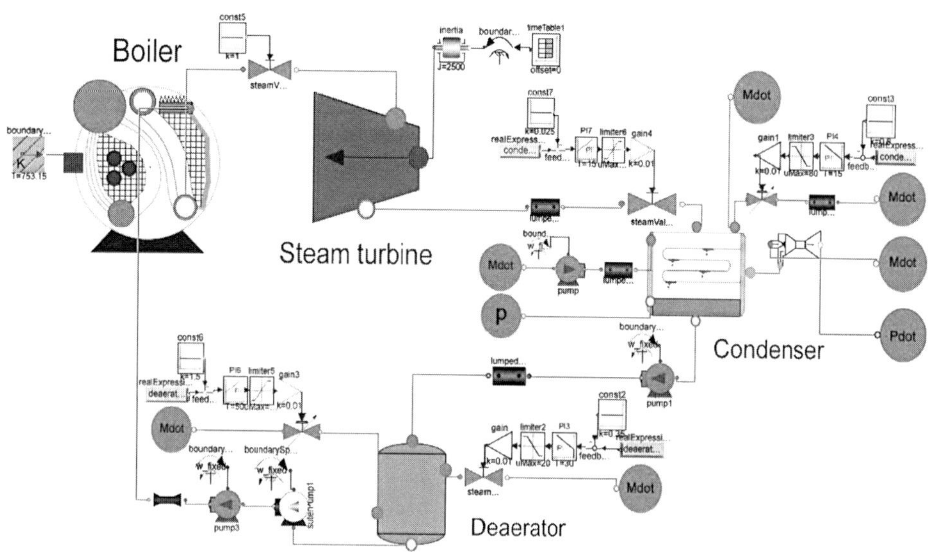

Figure 4. *Digital model of marine steam power system.*

where:

$\tilde{y}_i$-The dimensionless parameter;

y_s-The simulation or experimental parameter;

y_d-The design parameter.

The steam turbine output power, main steam flow, main steam pressure, main steam temperature, and cooling water outlet and inlet temperatures are shown in Figures 5-10.

It can be seen from the load increase test data for the marine steam turbine that the steam turbine reached a 100% working condition in five steps after starting up. After the system warmed up for 650 s, the flow of steam through steam valve was gradually increased, and after 270 s, the load was increased to about 20% of the working condition, and then the system operated stably for 740 s. Subsequently, the steam flow was increased again. After 190 s, the system rose to about 40% of the working condition and ran stably for 270 s under the 40% working condition. Then, an additional increase of the steam flow followed. After 180 s, the system rose to about 60% of the working condition. When the system had operated for 450 s under the 60% working condition, the steam flow was increased. After 300 s, the system rose to about 80% of the working condition, and then the system operated stably for 320 s. A final increase of the steam flow followed. After 200 s, the system rose to a 100% working condition. Under the condition that the main steam temperature, pressure and cooling water flow are kept under full working condition parameters, the load of the steam turbine unit can be increased by adjusting the main steam flow. The main steam temperature and pressure fluctuate due to various factors during operation. However, the state of fluctuation is controllable and safe.

The measured boundary parameters of the turbine inlet are substituted into the established digital model of the marine steam power system, and a simulation

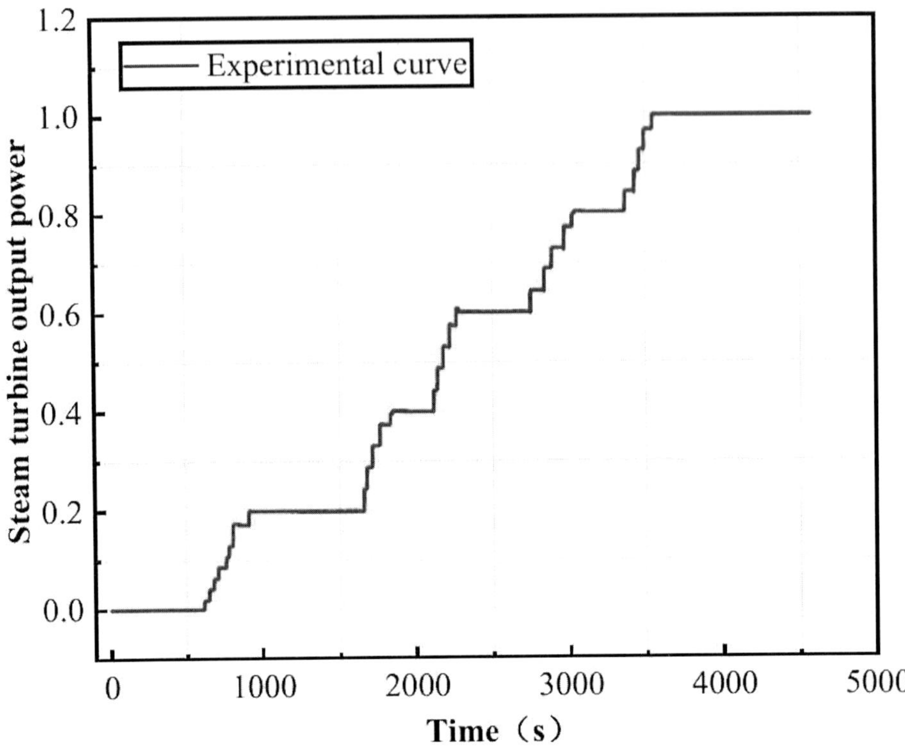

Figure 5. *Steam turbine output power.*

Table 1. *Comparison between experimental and simulation data of main steam flow at different operating points.*

Main Steam Flow	Experimental Data	Simulation Data	Relative Error
T = 800 s	0.109	0.104	−4.5%
T = 1800 s	0.345	0.328	−4.9%
T = 3000 s	0.706	0.716	1.4%
T = 4000 s	0.954	0.968	1.5%

curve for the turbine under 100% load rise conditions after start-up is obtained. The operation state parameters for key pieces of thermal equipment are compared with the digital simulation model for verification. The comparison diagram for steam turbine output power and the dynamic characteristic curve for the increasing temperature of the cooling water are shown in Figures 11-13.

It can be seen from Figure 11 that the simulation data for the main steam flow are consistent with the experimental data, but that, with the increase of the steam flow, the parameters fluctuate greatly as the load increases. Therefore, four operating points from the experimental and simulation data are selected for error analysis, as shown in Table 1.

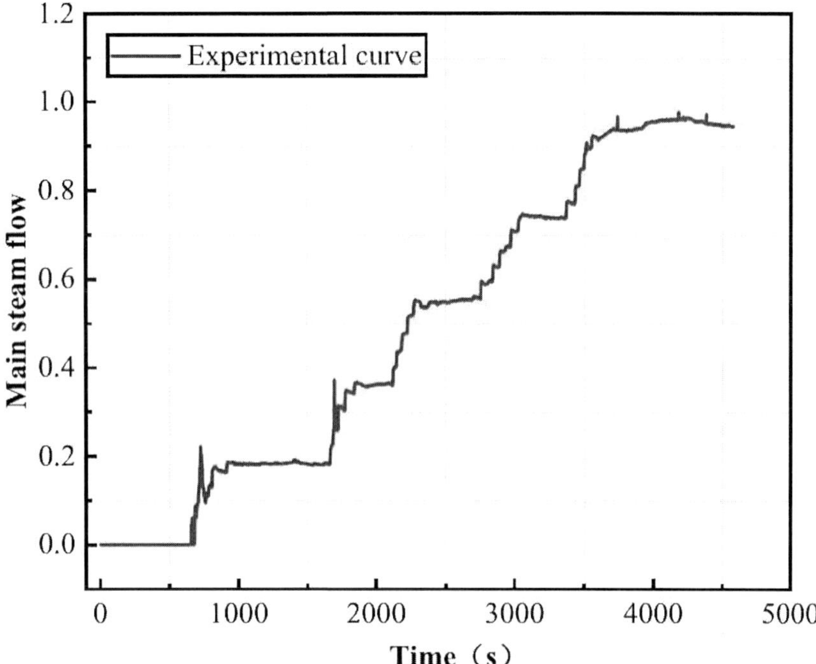

Figure 6. *Main steam flow.*

Table 2. *Comparison between experimental and simulation data of steam turbine output power at different operating points.*

Steam Turbine Output Power	Experimental Data	Simulation Data	Relative Error
T = 800 s	0.120	0.121	0.8%
T = 1800 s	0.372	0.356	−4.3%
T = 3000 s	0.771	0.766	−0.6%
T = 4000 s	0.995	0.994	−0.1%

It can be seen from the Table 1 that, among the four selected operating points, the one with the largest error, −4.9%, is T = 1800 s. This is because, in the process of lifting the load, the steam flow fluctuates easily. The error at the other operating points does not exceed ±5%.

It can be seen from Figure 12 that the simulation data for the steam turbine output power are consistent with the experimental data, but that, with the increase of the steam flow, the parameters fluctuate greatly during load increase. Therefore, four operating points from the experimental and simulation data are selected for error analysis, as shown in Table 2.

It can be seen from the Table 2 that, among the four selected operating points, the one with the largest error, −4.3%, is T = 1800 s, and that the error at other operating points does not exceed ±5%.

It can be seen from Figure 13 that the simulation data for the cooling water outlet temperature are consistent with the experimental data, but that, with the

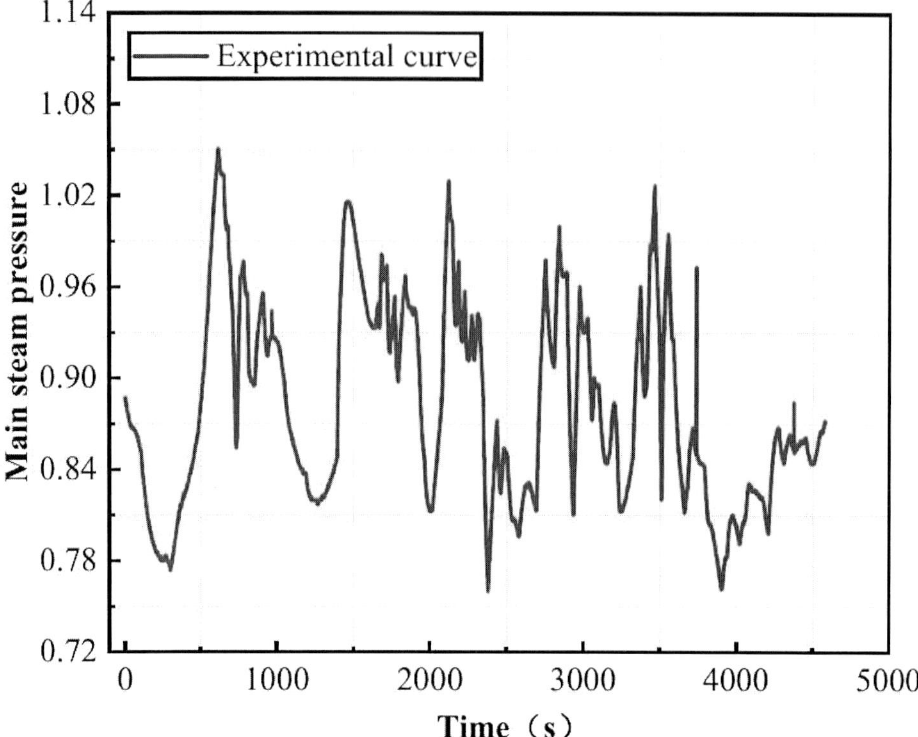

Figure 7. *Main steam pressure.*

Table 3. *Comparison between experimental and simulation data of cooling water outlet temperature at different operating points.*

Cooling Water Outlet Temperature	Experimental Data	Simulation Data	Relative Error
T = 800 s	0.749	0.753	0.5%
T = 1800 s	0.787	0.799	1.5%
T = 3000 s	0.885	0.891	0.7%
T = 4000 s	0.978	0.991	1.3%

increase of the steam flow, the parameters fluctuate greatly as the load increases. Therefore, four operating points from the experimental and simulation data are selected for error analysis, as shown in Table 3.

It can be seen from Table 3 that, among the four selected operating points, the one with the largest error, 1.5%, is T = 1800 s, and that the error of other operating points does not exceed ±5%.

Through the comparison and error analysis of the simulation curve and the experimental curve, it can be concluded that the main steam flow, the steam turbine output power and the cooling water outlet temperatures, etc. calculated by the digital model are close to the experimental curve, the dynamic operation trend is

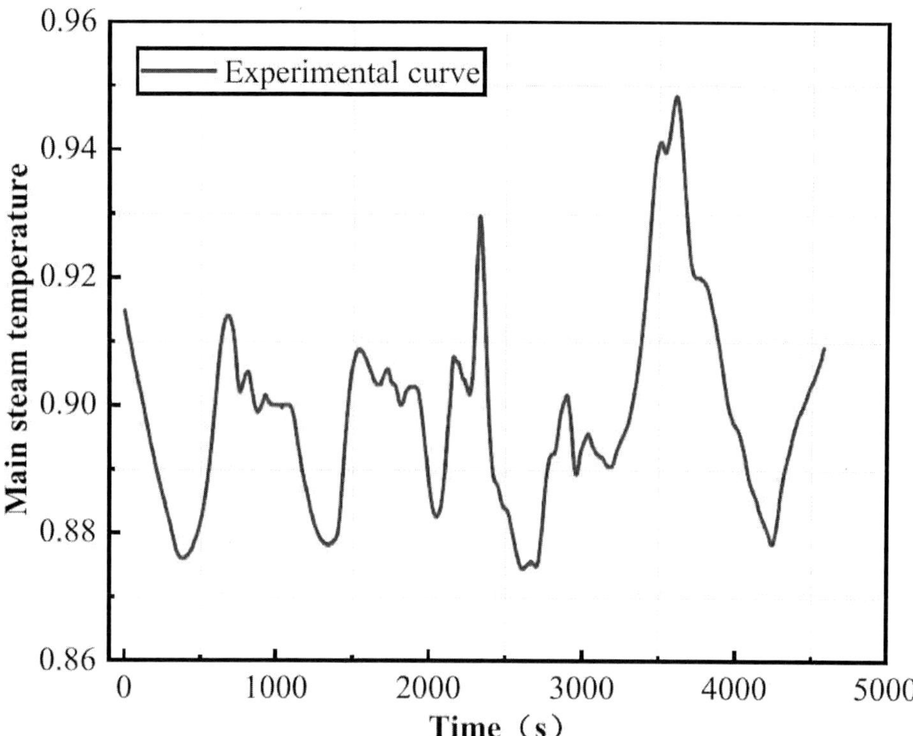

Figure 8. *Main steam temperature.*

consistent, and the maximum error does not exceed ±5%, which indicates a good simulation of the dynamic change process of the steam turbine. The digital model established in this paper has high accuracy and can realize virtual reality mapping between the model and the actual equipment. It can be used for the dynamic analysis of marine steam power systems and is conducive to promoting the further application of digital twin technology.

4. DYNAMIC CHARACTERISTICS ANALYSIS OF MARINE POWER SYSTEM BASED ON DIGITAL MODEL

Maneuverability is an important performance indicator for marine technology. In the process of changing the marine vessel's navigational maneuverability, the power system will rapidly increase or reduce the load, and the boiler, main steam turbine and steam water circulation system will coordinate to control the load change, which is very likely to cause a mismatch in turbine-boiler coordinated control, an imbalance in the steam water system, large fluctuations in the auxiliary system parameters and other phenomena [16].

However, the fast load changing capability of the marine steam power system is an important factor in determine the maneuverability of the marine vessel, and

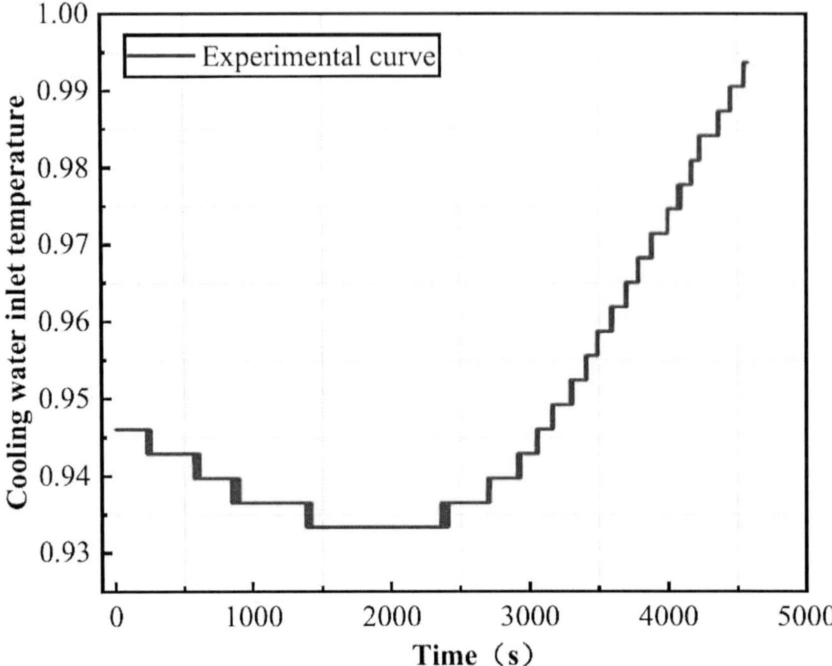

Figure 9. *Cooling water inlet temperature.*

its dynamic response performance will directly affect the operational stability of the marine power system. While mastering the dynamic characteristics of its operation can provide guidance for the vessel's operation strategy, this section uses the high-precision digital model of the marine steam power system after the steady-state verification to analyze its dynamic characteristics, focusing on the dynamic response of the system under different cooling conditions and the dynamic characteristics under lifting and lowering loads.

4.1. Analysis of Dynamic Characteristics of System under Lifting and Lowering Loads

In order to accurately grasp the dynamic characteristics of the operation of the marine steam power system, this section analyzes the dynamic characteristics of the digital model of the system under variable working conditions and simulates the operating characteristics of the system under variable working conditions at 1500 s.

4.1.1. Dynamic Characteristic Analysis of Marine Power System Load Decrease

As can be seen in Figures 14-17, when the working condition of the steam turbine changes from 100% to 70%, the turbine's steam consumption is reduced due to the reduction of the load of the steam turbine. The reduction in steam flow

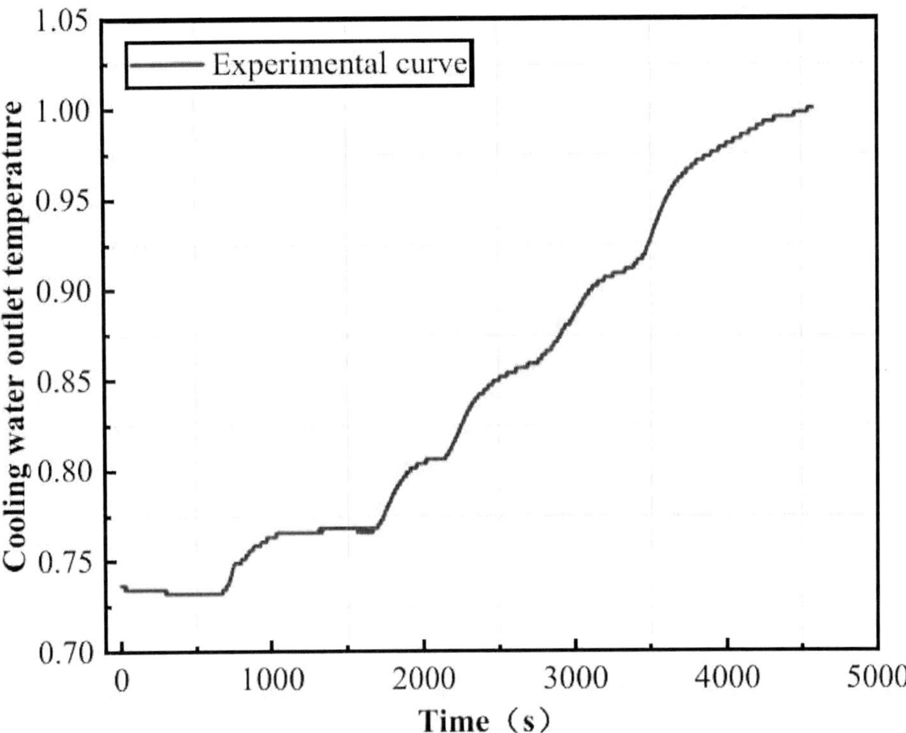

Figure 10. *Cooling water outlet temperature.*

will result in a significant reduction in the amount steam in the condenser, thus reducing the water level in the condenser. As the flow of cooling water is sufficient, the pressure in the condenser will also be reduced, so the vacuum degree of the condenser will rise rapidly in a short time. As the amount of superheated steam decreases, the water level in the boiler will rise. As a result, the flow of feed water passing through the deaerator into the boiler will be reduced, leading to the increase in the water level inside of the deaerator.

4.1.2. Dynamic Characteristic Analysis of Marine Power System Load Increase

As can be seen from Figures 18-21, when the working condition of the steam turbine changes from 50% to 70%, the steam consumption of the turbine increases and the flow of cooling water sufficient to absorb the heat from the steam turbine exhaust, so the pressure in the condenser increases, and the water level in the condenser will also rise with the increase of the condensed steam. The increasing amount the superheated steam will also lead to the decrease in the water level in the boiler. As a result, the flow of feed water from the deaerator into the boiler will increase, leading to a decrease in the water level in the deaerator.

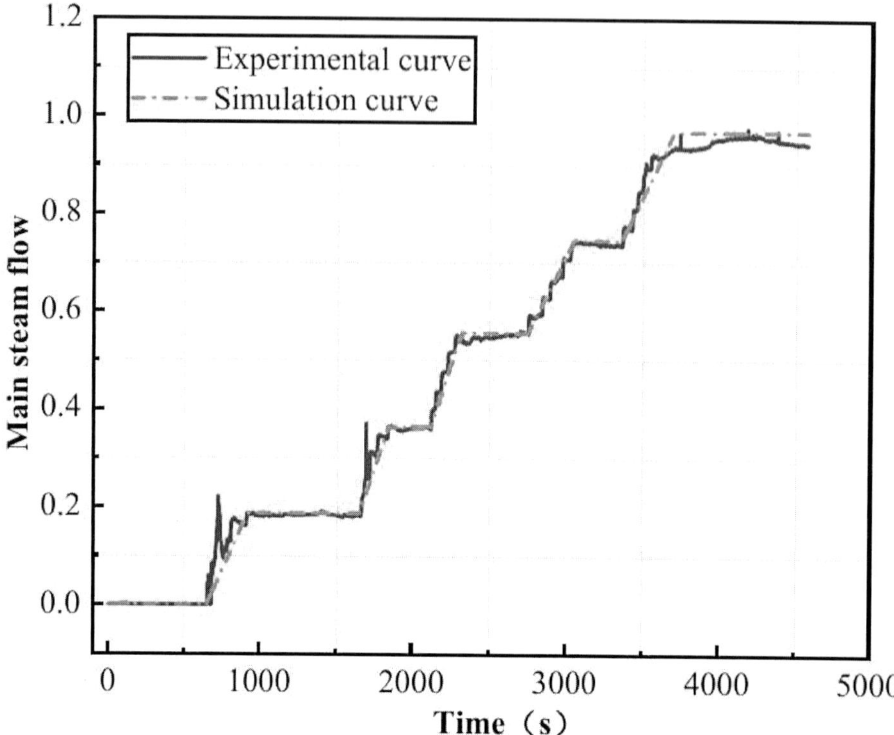

Figure 11. *Comparison between experimental and simulation data of main steam flow.*

Table 4. *Influence of different cooling water flow on the system.*

Cooling Water Inlet Temperature (°C)	Relative Flow of Cooling Water	Condenser Pressure Increment (Bar)	Power Increment of Steam Turbine (MW)
	0.6	0.086	−1.584
	0.7	0.053	−1.057
	0.8	0.030	−0.65
	0.9	0.013	−0.349
25	1.0	0	0
	1.1	−0.011	0.311
	1.2	−0.021	0.51
	1.3	−0.030	0.735
	1.4	−0.037	0.936

4.2. Dynamic Response of the System under Different Cooling Conditions

(1) Increase or decrease of cooling water flow

In order to master the influence of different cooling water flows on the steam power system, the digital model of the marine steam power system is used to analyze the changes in its parameters for different flow rates of the cooling water. The changes in the parameters are shown in Table 4.

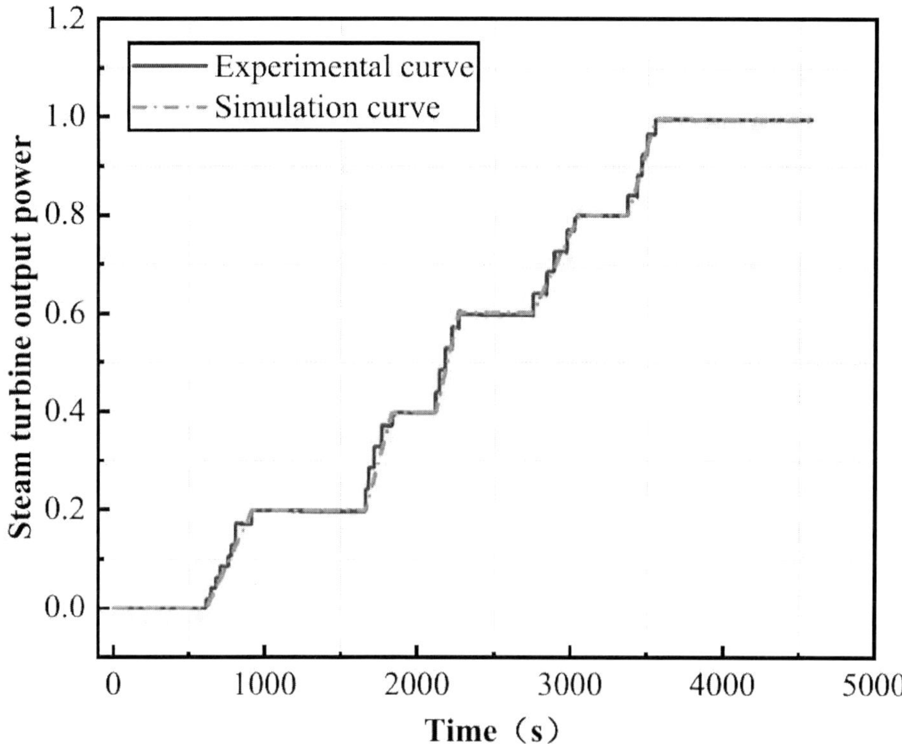

Figure 12. *Comparison between experimental and simulation data of steam turbine output power.*

It can be seen from Table 4 that, when the inlet temperature of the cooling water is constant, the condenser pressure decreases with the increase of the cooling water flow and increases with the decrease of the cooling water flow. The power increment of the steam turbine increases with the increase of the cooling water flow and decreases with the decrease of the cooling water flow.

(2) Increase or decrease in cooling water temperature

In order to master the influence of different cooling water temperatures on the steam power system, the digital model of the marine steam power system is used to analyze the changes in the parameters at different inlet temperature of the cooling water. The changes in the parameters are shown in the Table 5.

It can be seen from Table 5 that, when the relative flow of cooling water is constant, the condenser pressure increases with the increase of the cooling water inlet temperature and decreases with the decrease of the cooling water inlet temperature. The power increment of steam turbine decreases with the increase of cooling water inlet temperature, and increases with the decrease of cooling water inlet temperature.

According to the state parameters of each piece of thermal equipment, the relationship curves between the cooling water flow, the temperature and the power

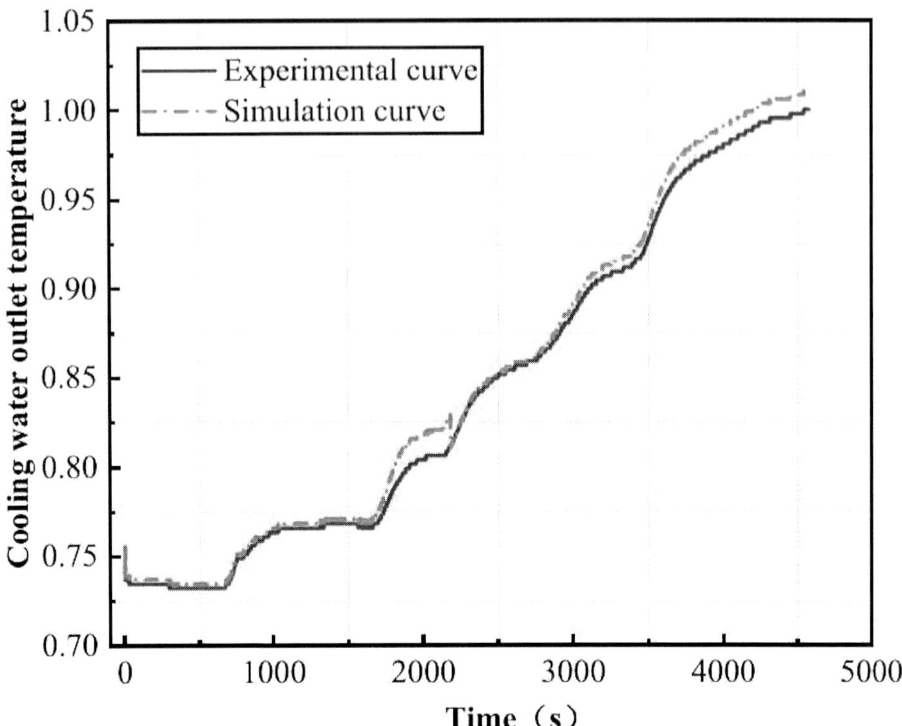

Figure 13. *Comparison between experimental and simulation data of cooling water outlet temperature.*

Table 5. *Influence of different cooling water temperature on the system.*

Relative Flow of Cooling Water	Cooling Water Inlet Temperature (°C)	Condenser Pressure Increment (Bar)	Power Increment of Steam Turbine (MW)
1.0	16	−0.042	1.075
	19	−0.031	0.788
	22	−0.017	0.411
	25	0	0
	28	0.021	−0.446
	31	0.045	−0.897

increment of the steam turbine, and the relationship curves between the cooling water flow, the temperature and the condenser pressure are drawn as shown in Figures 22 and 23.

It can be seen from Figure 22 that the relation between the cooling water flow and the condenser pressure increment is as follows:

(1) When the relative flow of cooling water gradually increases, it can be seen from the heat transfer calculation formula that the total heat transfer coefficient of the condenser will increase, the heat transfer end difference will decrease, and

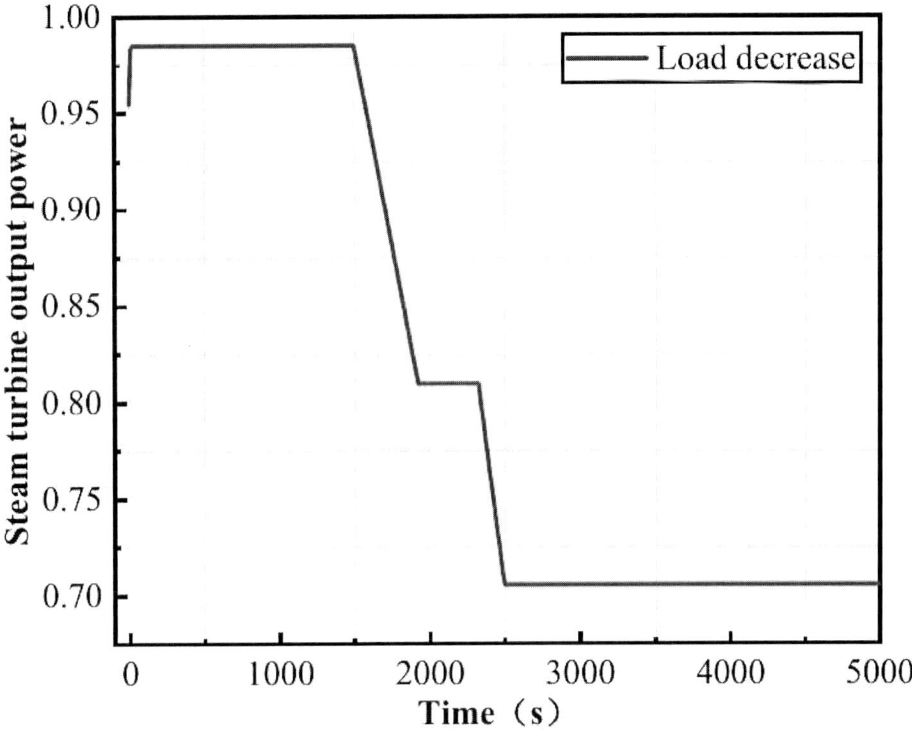

Figure 14. *The steam turbine output power under load decrease condition.*

the heat transfer performance of the condenser will improve, so that the condenser pressure will decrease, the vacuum degree will increase, and the heat cycle efficiency will increase, thus increasing the power of the main turbine. However, with the increase of the flow, the change range of the condenser pressure will gradually slow down [17];

(2) When the relative flow of the circulating cooling water decreases gradually, the cooling water flow rate also decreases, the total heat transfer coefficient of the condenser decreases, the heat transfer end difference increases, the heat transfer performance of the condenser decreases, the condenser pressure increases gradually, the vac- uum degree decreases, the thermal cycle efficiency decreases, and the steam turbine power decreases.

It can be seen from Figure 23 that the relation between cooling water inlet temperature and condenser pressure increment is as follows:

(1) When the inlet temperature of the circulating cooling water increases gradually, the heat exchange capacity of the cooling water decreases, the pressure in the condenser increases gradually, the vacuum degree decreases, the thermal cycle efficiency decreases, and the power of steam turbine decreases accordingly;

(2) When the inlet temperature of the circulating cooling water decreases gradually, the heat exchange capacity of the cooling water increases, the pressure in

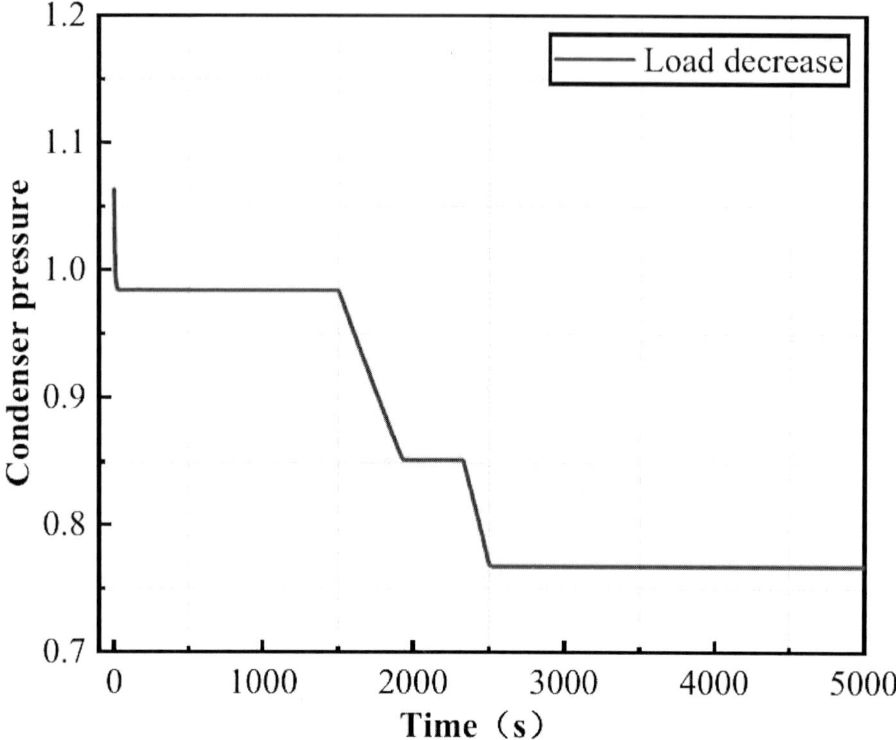

Figure 15. *The condenser pressure under load decrease condition.*

the condenser decreases gradually, the vacuum degree increases, the thermal cycle efficiency increases, and the power of the steam turbine increases accordingly.

Through the dynamic response of the digital model under different cooling conditions, it can be concluded that increasing the flow of the circulating cooling water can improve the heat transfer performance of the condenser, thereby reducing the condenser pressure, improving the condenser vacuum degree and the steam power cycle efficiency, and thus increasing the power of the main turbine. With the gradual increase of the cooling water flow, the condenser pressure and the power of the main turbine are still limited, and increasing the cooling water flow will also increase the output power of the circulating water pump. If the horsepower of the water pump increases, the power consumed is greater than the increment of the power of the main turbine, which is often not worth the loss. Therefore, when the marine vessel is performing tasks, the flow of the circulating cooling water should be dynamically adjusted according to the sea water temperature to ensure that the condenser maintains the best vacuum degree and maximizes the efficiency of the turbine.

In conclusion, the dynamic response of the system under different cooling conditions is analyzed using the digital model of the marine steam power system. The results show that the model can accurately reflect changes in the system's state

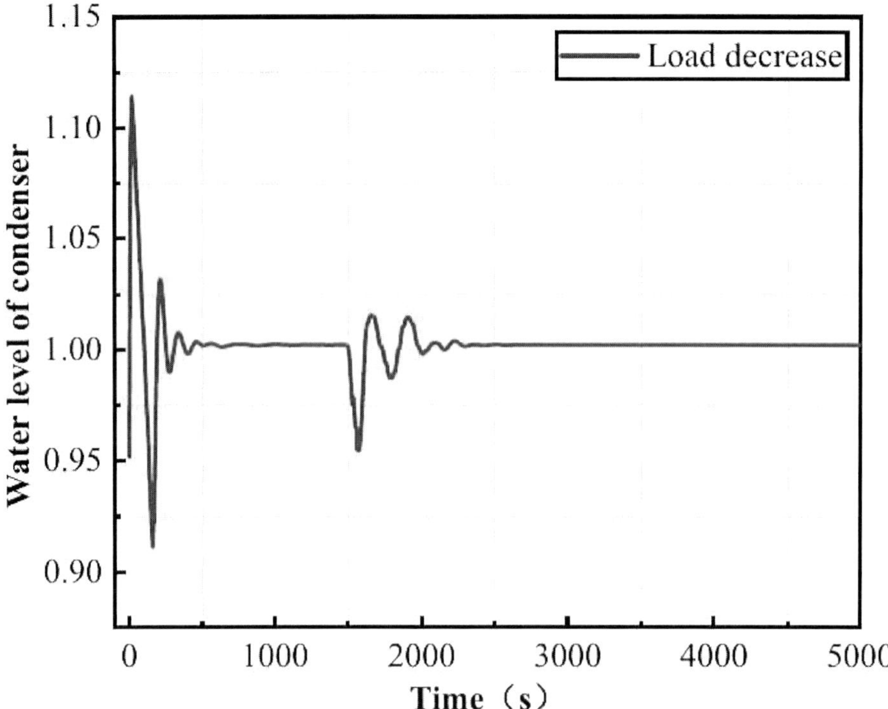

Figure 16. *The water level of condenser under load decrease condition.*

parameters, realize the virtual reality mapping between the digital model and the physical equipment, and provide guidance for the safe and stable operation of the marine steam power system under different cooling conditions.

5. FAULT SIMULATION ANALYSIS OF MARINE STEAM POWER SYSTEM BASED ON DIGITAL MODEL

A large number of data signals, such as flow, temperature, pressure, power and speed, are generated during the operation of the marine steam power system. These data signals reflect the operating characteristics and performance degradation laws of the thermal equipment. Nonetheless, it is still difficult to obtain data for each fault sample in the whole life cycle of the equipment. Especially for the multi-domain coupled marine system, due to the lack of fault sample data, it is not possible to accurately analyze the degradation law of equipment performance, nor to accurately achieve fault diagnosis and prediction.

However, fault simulation and prediction under the framework of digital twin technology can overcome the problem of obtaining fault sample data. It builds a high-precision digital simulation model of physical entities, optimizes and solves the simulation model by relying on online monitoring, data analysis, fault prediction, life-cycle management, etc., and feeds back the simulation results to physical

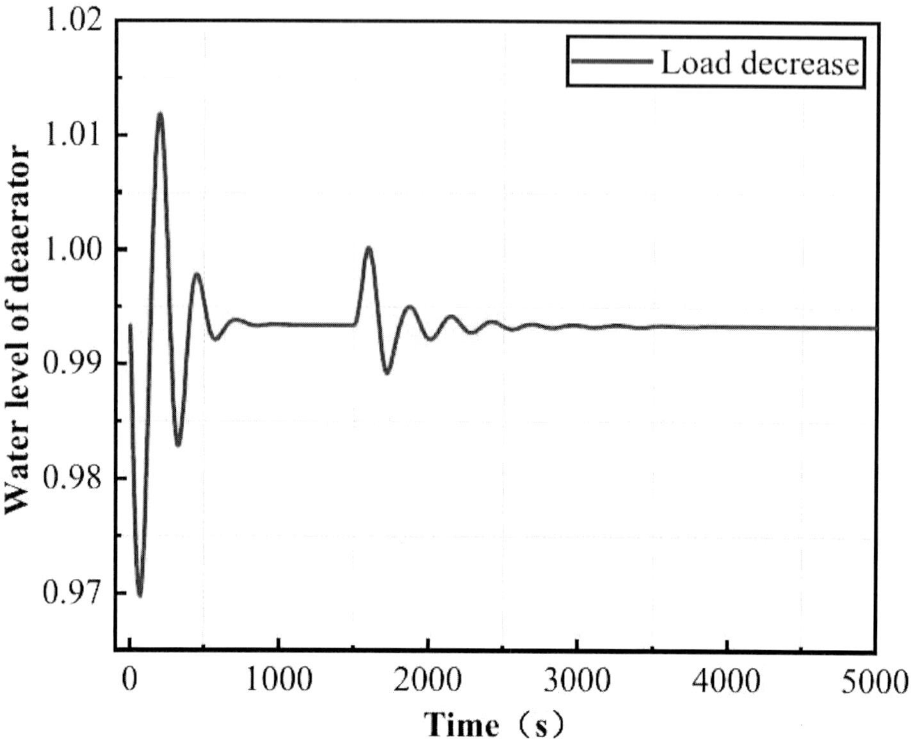

Figure 17. *The water level of deaerator under load decrease condition.*

objects, so as to optimize and make decisions for physical objects. Relying on digital twin technology, fault simulation and prediction based on high-precision digital models can effectively realize the reverse mapping of digital models to physical equipment. The digital model can train the model using the fault sample data to lay the foundation for subsequent fault diagnoses. It can also inject fault points, and, through the dynamic analysis of fault behavior, grasp the law of performance degradation. The fault diagnosis process is shown in Figure 24.

As can be seen from Figure 24, which shows the fault diagnosis process for the marine steam power system based on a digital model, using the digital model to simulate equipment faults and extract fault data is the most important step in establishing a fault database and is also the basis for realizing fault diagnoses. Therefore, this section takes the marine condensate feed water system as an example, and, making full use of the advantages of its digital model, inserts fault points on the basis of the digital model, conducts a quantitative analysis of the main fault types of the thermal equipment in the condensate feed water system, and uses the model to accurately describe the actual fault status, so as to realize the monitoring and early warning of faults by the digital model. On the one hand, this provides targeted measures to guide the safe and stable operation of marine vessel; on the

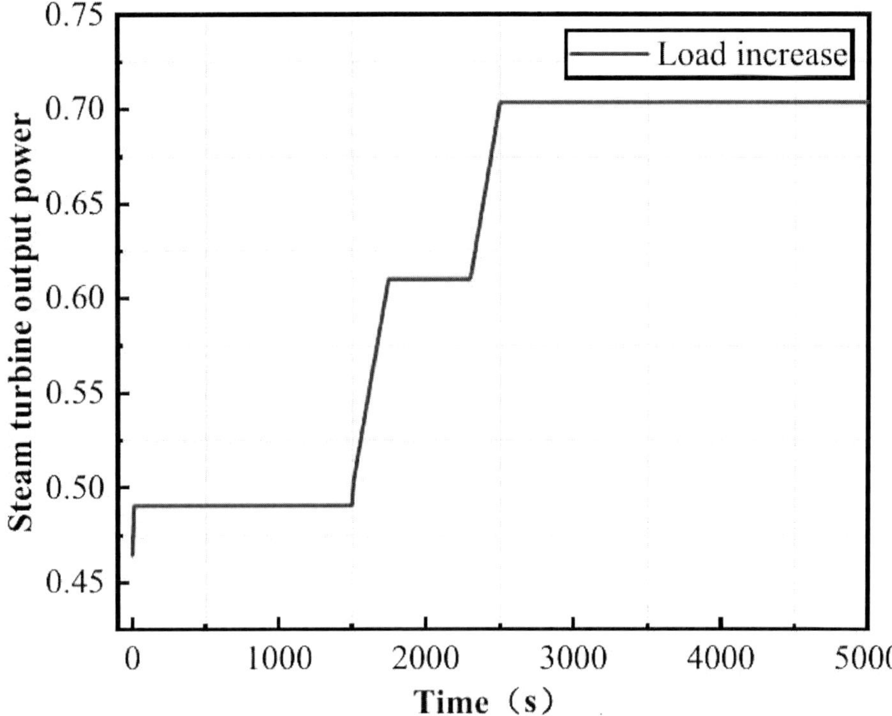

Figure 18. *The steam turbine output power under load increase condition.*

other hand, it realizes the reverse mapping of digital models to physical equipment, which is also one of the important components of digital twin technology.

As an important auxiliary system in the marine steam power system, the condensate feed water system and its operational reliability directly affect the whole power system. Influenced by complexity, randomness, dynamics and other factors, the condensate feed water system will also encounter various abnormal conditions in its operation. For each type of thermal equipment failure, accurate and effective measures need to be taken to regulate it. If this is not handled properly, frequent accidents will occur, causing equipment damage or unit shutdown. According to thermal calculations and on-site operating experience, the main causes of the failure of the condensate feed water system are water level regulation failure, the vacuum system not being tight, abnormal operation of the steam jet ejector, failure of the circulating water pump, reduction of the circulating water flow, a rise in the temperature of the cooling water, blockage of a cooling pipe, excessive steam exhaust in the steam turbine, etc. [18]. The main faults of the condensate feed water system are systematically analyzed below.

(1) Failure of water level regulation

The regulation of the water levels in the condenser and deaerator, which plays an important role in the safe and stable operation of the condensate feed water system, mainly involves the marine condensate feed water system. If the water level

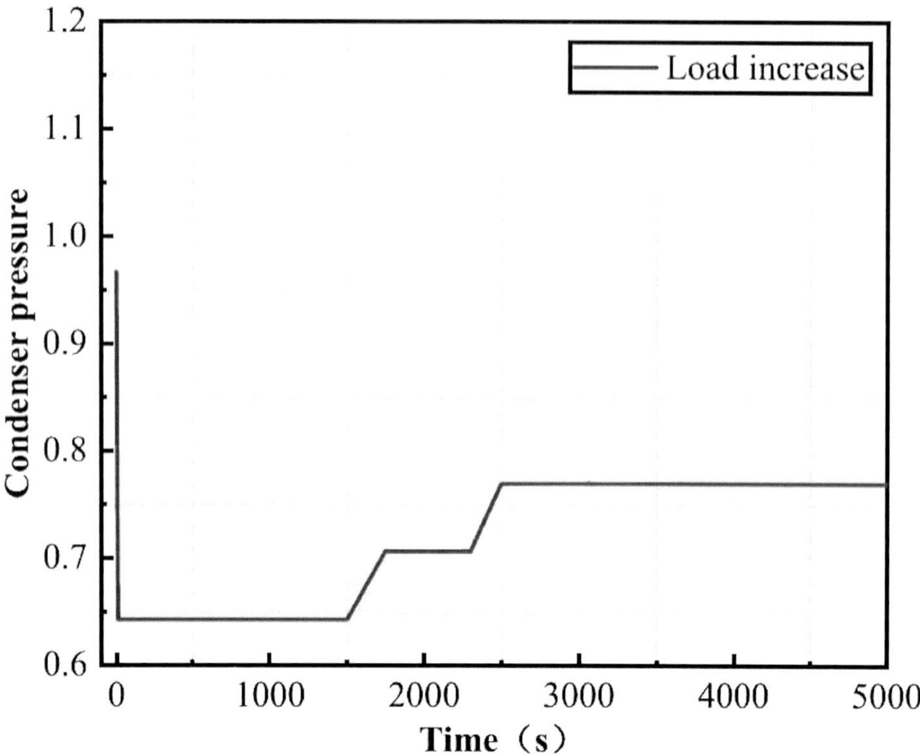

Figure 19. *The condenser pressure under load increase condition.*

in the condenser is too high, the cooling effect will be weakened and the vacuum degree of the unit will drop, and, if the water level is too low, cavitation of the condensate pump will occur, affecting the safety of the unit. Too high a water level in the deaerator will cause unstable pressure in the deaerator and affect the deaerating effect. Too low a water level will reduce the inlet pressure of the feed pump and cause cavitation of the pump, which will damage the feed pump in serious cases. Therefore, based on the digital model, a fault simulation study is carried out on the water level regulation failure of the condensate feed water system, and a quantitative analysis is made on the change rule of the main parameters of the system caused by the water level regulation failure. The influence of water level regulation failure on the main state parameters of the condensate feed water system is shown in Figures 25-28.

Through a quantitative simulation study of the failure of water level regulation in the condensate feed water system, it can be seen that, when the water level regulation is sluggish, the water level in the condenser and deaerator fluctuates continuously, thus causing the fluctuation of the condenser pressure and the turbine output power. If the water level regulation is sluggish for a longer time, or even fails, the safety and stability of the condensate feed water system will be greatly

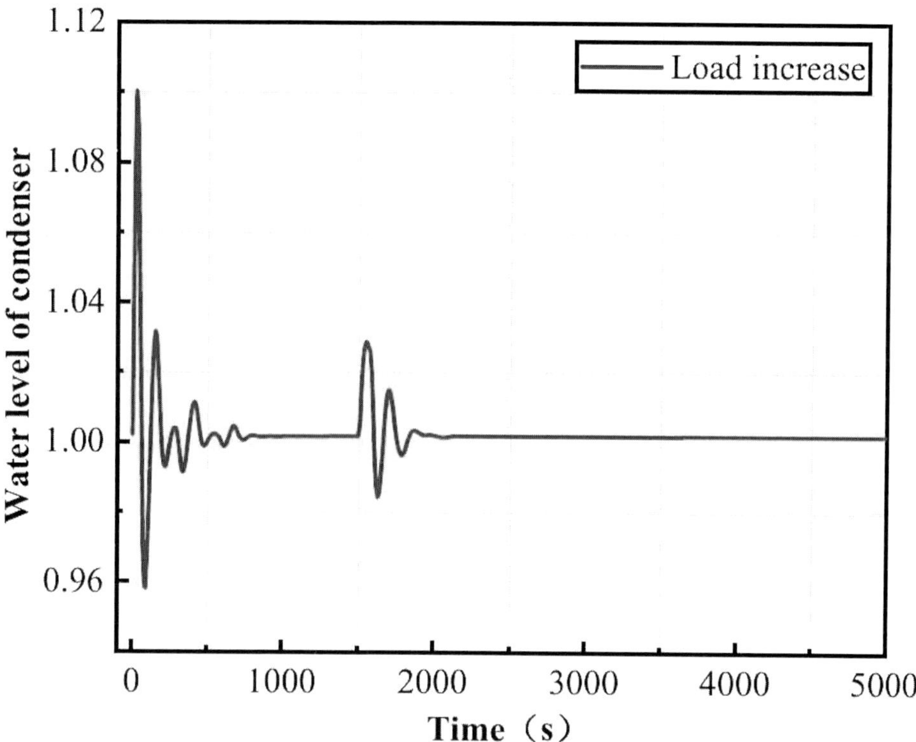

Figure 20. *The water level of condenser under load increase condition.*

affected. Therefore, attention must be paid to water level regulation during steam turbine operation to avoid failure.

(2) Air leakage of the condenser vacuum system

During the operation of the condenser, air will inevitably leak in. It is precisely because the air extractor can draw out non-condensable gas from the condenser that the condenser can maintain a good vacuum. When the working state of the air extractor remains unchanged and the amount of air leaking in increases, the proportion of air in the condenser will increase.

In order to analyze the influence of different amounts of air leakage into the condenser on the condenser, take a 70% working condition as an example, assume that other input and output variables of the system remain unchanged, and simulate and analyze the changes of the main operating state parameters of the condenser when the air leakage increases from 10% to 60%. The influence of the increased air leakage on the main state parameters of the condenser is shown in Table 6.

According to the state parameters of the condenser, the relationship curves between the pressure increment of the condenser, the temperature increment of the circulating cooling water and the increase proportion of the air leakage are drawn as shown in Figures 29 and 30.

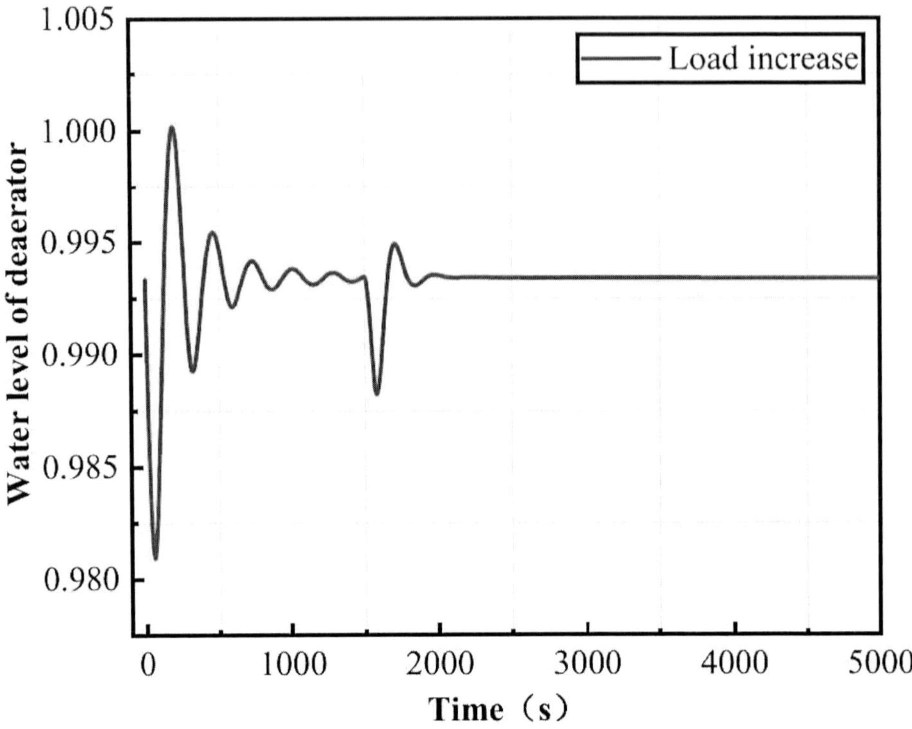

Figure 21. *The water level of deaerator under load increase condition.*

Table 6. *Influence of increased air leakage on main state parameters of condenser.*

Increase Proportion of Air Leakage (%)	Condenser Pressure Increment (Bar)	Temperature Rise Increment of Cooling Water (°C)
10	0.011	−0.11
20	0.023	−0.23
30	0.035	−0.42
40	0.052	−0.71
50	0.076	−1.05
60	0.113	−1.38

It can be seen from Figures 29 and 30 that, with the increase of air leakage into the condenser, the proportion of the air volume inside the condenser increases. According to Dalton's partial pressure law, the pressure in the condenser will also increase, reducing the thermal cycle efficiency of the turbine. In addition, the air film formed will reduce the heat transfer coefficient, and the heat transfer end difference of the condenser will also increase.

(3) Increase of steam turbine exhaust

During the steady-state operation of the steam turbine unit, the cooling conditions of condenser also remain unchanged. When the exhaust steam in the steam

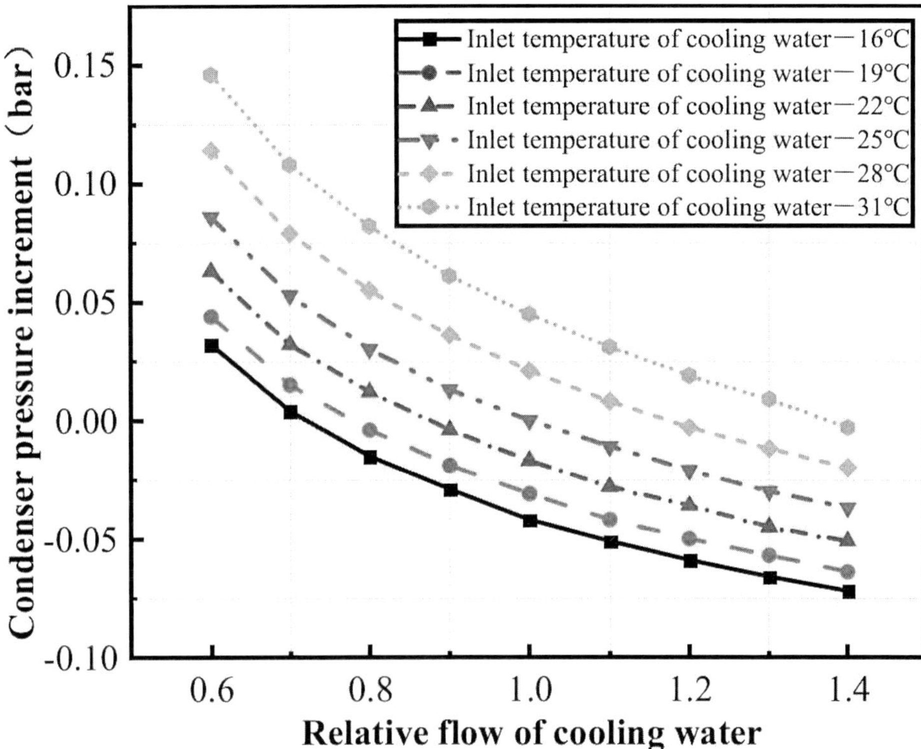

Figure 22. *Relation curve between cooling water flow, temperature and condenser pressure increment.*

Table 7. *Influence of increased steam turbine exhaust on main state parameters of condenser.*

Increase Proportion of Steam Turbine Exhaust (%)	Condenser Pressure Increment (bar)	Temperature Rise Increment of Cooling Water (°C)
5	0.015	0.46
10	0.029	0.77
15	0.048	1.25
20	0.062	1.84
25	0.098	2.31

turbine increases, the steam quantity inside the condenser will change. In order to study the rule of the influence of steam turbine exhaust on the condenser, assuming that other input and output variables of the system remain unchanged. The change in the main operating state parameters of the condenser due to the increase of steam turbine exhaust is analyzed according to the established digital model. The main performance parameters of the condenser are shown in Table 7.

According to the state parameters of the condenser, the relationship curves between the pressure increment of the condenser, the temperature increment of the circulating cooling water and the increased proportion of the steam turbine exhaust are drawn as shown in Figures 31 and 32.

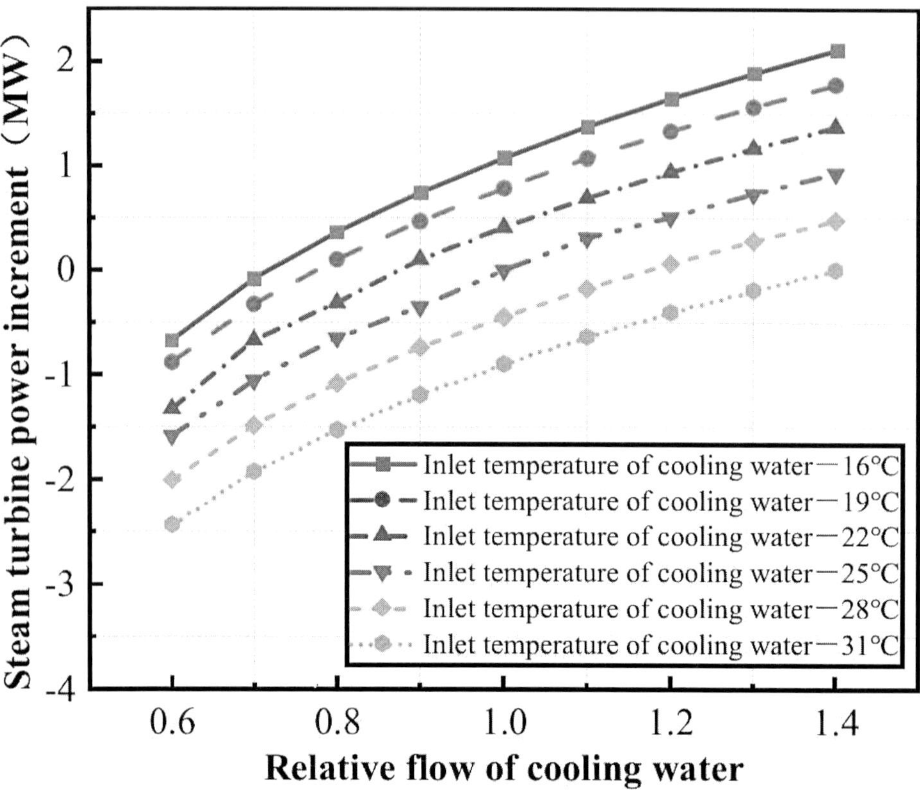

Figure 23. *Relation curve between cooling water flow, temperature and steam turbine power increment.*

It can be seen from Figures 31 and 32 that, if the cooling conditions of the condenser remain unchanged, when the steam exhaust volume of the turbine gradually increases, the proportion of steam in the condenser increases, the pressure of the condenser rises, the vacuum degree drops, the cooling water temperature rises, and the heat transfer end difference of the condenser also increases, which is consistent with the mechanism model analysis.

In a word, through the fault simulation research on the condensate feed water system based on the high-precision digital model, it can be seen from the fault simulation results that the digital model can accurately simulate the change law for each fault parameter, accurately describe the actual fault state, realize the quantitative analysis of the fault using the digital model, and build the fault database of the digital twin model according to the parameter change laws for these faults. When the actual digital twin model is running, it can be compared in time through the real-time mapping of online data to extract typical faults, which can speed up online diagnosis and fault prediction and provide guidance for the safe and stable operation of the marine steam power system.

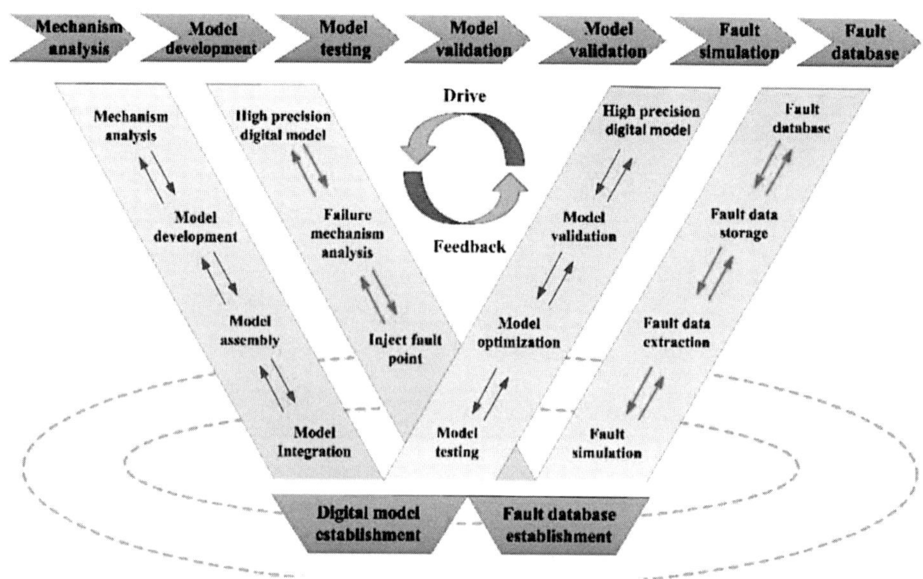

Figure 24. *Fault diagnosis process of marine power system based on digital model.*

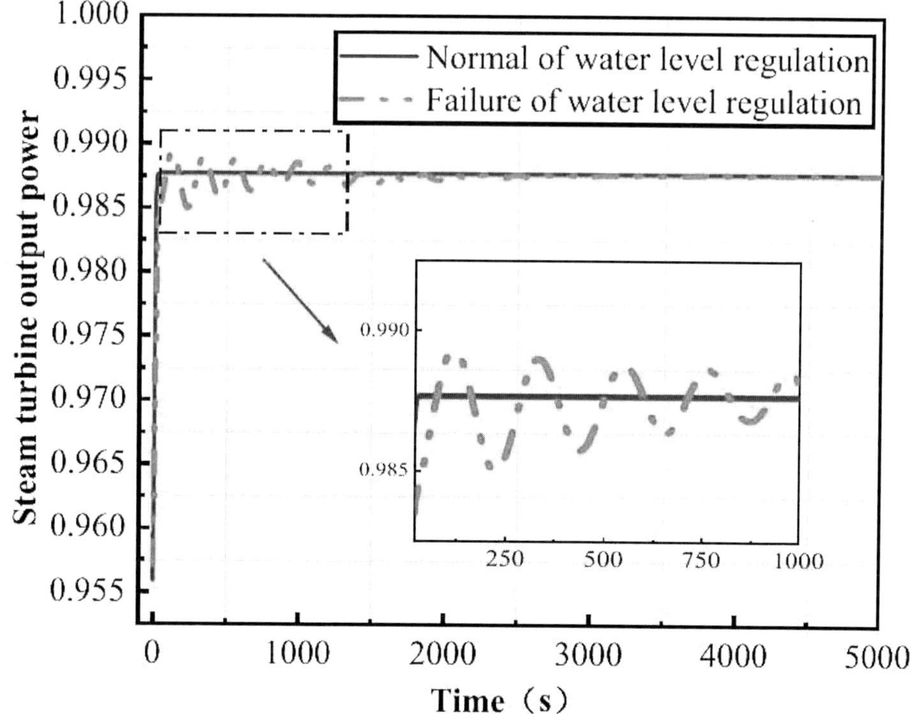

Figure 25. *The steam turbine output power under normal and failure of water level regulation.*

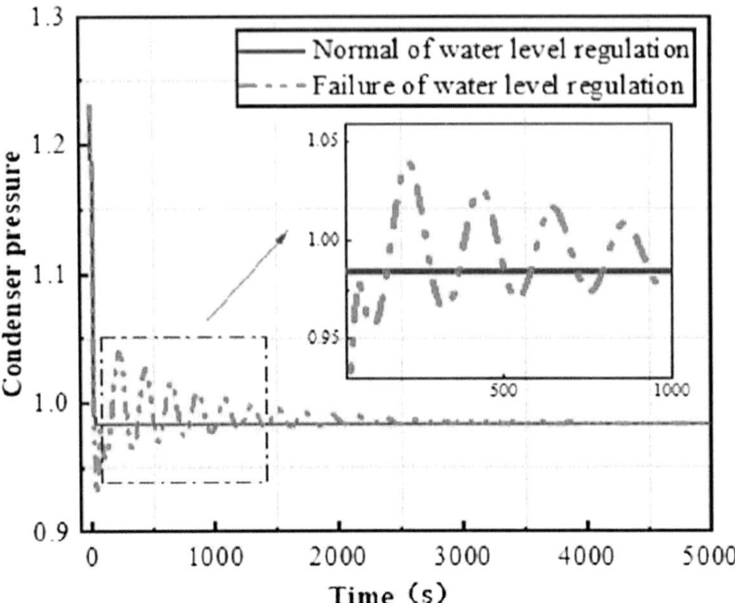

Figure 26. *The condenser pressure under normal and failure of water level regulation.*

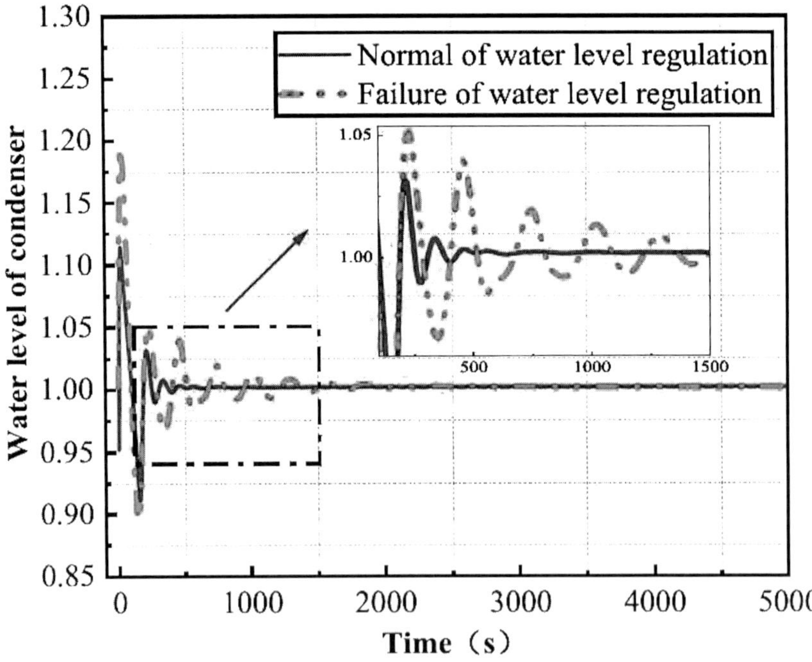

Figure 27. *The water level of condenser under normal and failure of water level regulation.*

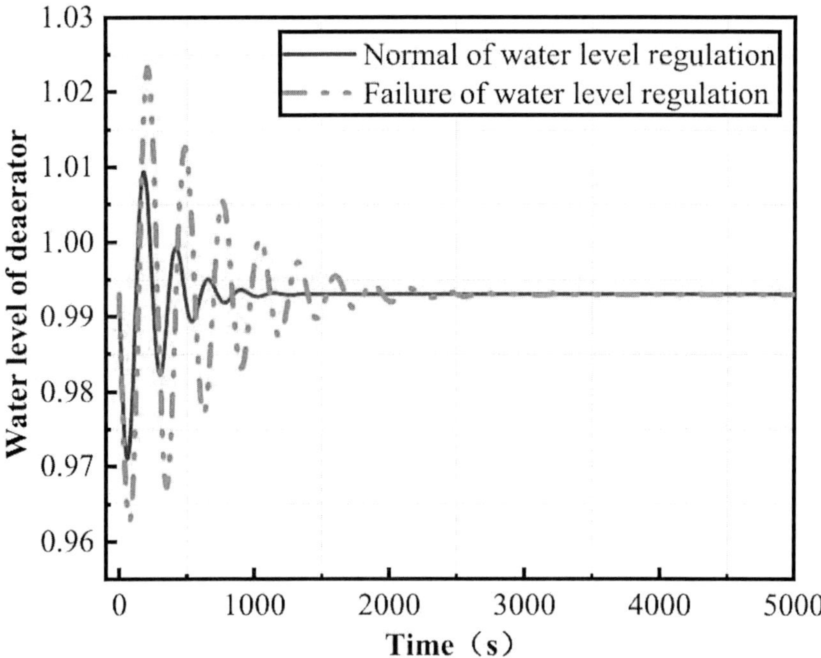

Figure 28. *The water level of deaerator under normal and failure of water level regulation.*

6. DISCUSSION

Using the digital twin technology system based on the Modelica language, combined with multi-domain physical modeling, modular modeling, etc., this paper conducts in- depth research on the multi-domain modeling, dynamic characteristic analysis and fault simulation of the marine steam power system.

(1) Using the digital twin technology system, the construction process of the digital model of the marine steam power system is given, including five steps: mechanism analysis, model development, model testing, model verification and model analysis. Based on its operational principle, a mathematical model of the equipment is established.

(2) Based on the multi-domain modeling language Modelica and the system simulation platform of MWorks, combined with the modular modeling, the equipment and subsystem model library is developed, the digital model of the marine steam power system is integrated, and the model characteristic parameters are debugged. Finally, the correctness and high accuracy of the model are verified using experimental data for a certain type of steam turbine, ensuring the accuracy of the digital twin model.

(3) Based on the digital model, the dynamic characteristics of the marine steam power system are analyzed. Through comparison with the test data, the error

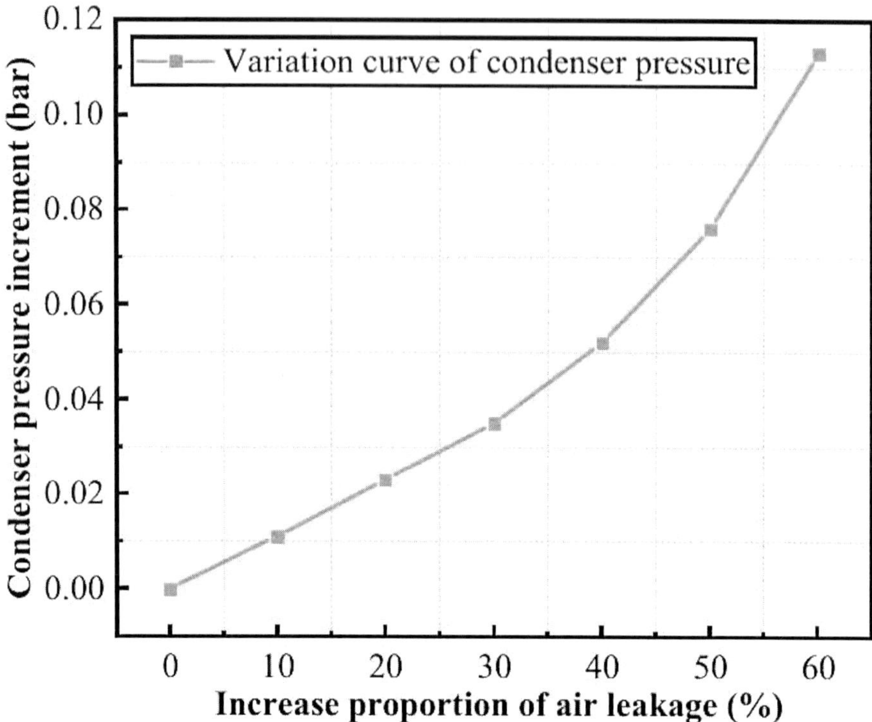

Figure 29. *Variation curve of condenser pressure.*

for the key performance parameters is established to be within ±5%. The system's dynamic change laws under different cooling conditions and load conditions are analyzed. The results show that the dynamic operation trend is consistent with the actual situation. The operation characteristics of the unit under variable working conditions are mastered, and virtual reality mapping between the digital model and the physical system is realized.

(4) Based on the digital model, the fault simulation of the marine steam power system is carried out, and a fault diagnosis process is proposed. Later, based on this digital model, a complete fault database for the marine steam power system will be established, and fault prediction for the marine steam power system based on digital twins will be carried out in subsequent research.

7. CONCLUSIONS

Aiming at the problems of the marine steam power system involving multi-disciplinary, multi-parameters, strong coupling, etc. characteristics, using the system of digital twin technology, combined with multi-disciplinary physical modeling and modular modeling, this paper has carried out in-depth research on its multi-disciplinary modeling, dynamic characteristics and fault diagnosis. The main contributions include:

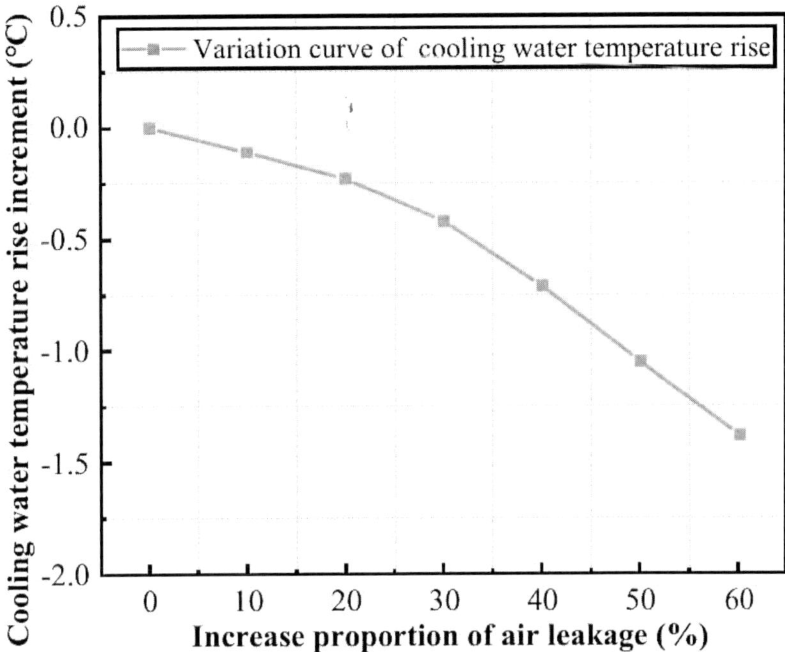

Figure 30. *Variation curve of cooling water temperature rise.*

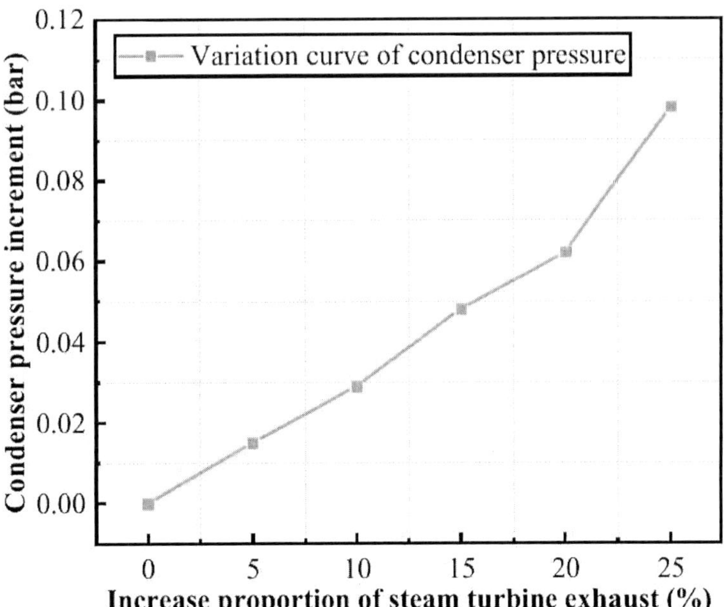

Figure 31. *Variation curve of condenser pressure.*

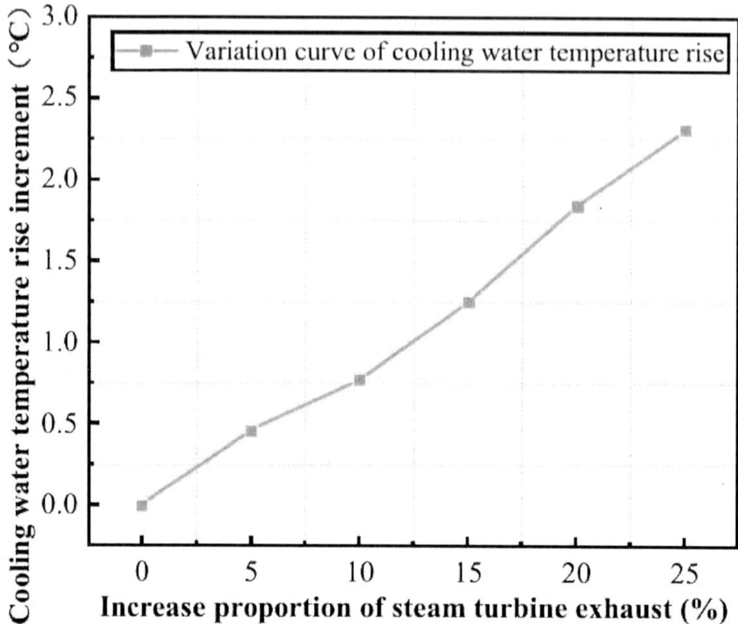

Figure 32. *Variation curve of cooling water temperature rise.*

(1) This paper introduces the problems and necessity of the dynamic characteristics of the marine steam power system, describes in detail the research progress in marine power system simulation and multi-domain modeling, conducts research on the multi-domain modeling of the marine steam power system, and presents the process of its digital model construction, including five steps: mechanism analysis, model development, model testing, model verification and model analysis.
(2) Based on the multi-domain modeling language Modelica and the system simulation platform of MWorks, combined with the modular modeling, a model library for the marine steam power system is developed, and a digital model of the marine steam power system is constructed. The model characteristic parameters are debugged using system operation data, and the high precision of the model is verified according to the experimental data for a certain type of steam turbine, ensuring the accuracy of the digital twin model.
(3) Based on the digital model, the dynamic characteristics of the marine steam power system are analyzed, and the dynamic characteristics of the system under variable working conditions are analyzed. The system operating state parameters under different cooling conditions and lifting and lowering system loads are analyzed. The analysis results show that the dynamic response trend of the model is consistent with the actual operation, and the error for the main steam flow, steam turbine output power, cooling water outlet temperature and other key parameters is within ±5%, meaning that the virtual

reality mapping between the digital model and the physical equipment has been realized.

(4) A fault diagnosis process for the marine steam power system based on the digital model is given. Taking the condensate feed water system as an example, the main faults of the thermal equipment in the condensate feed water system are analyzed based on the digital model. The main state parameters of the condensate feed water system are quantitatively analyzed for conditions resulting from water level control system failure, air leakage in the condenser vacuum system and exhaust increase in the steam turbine. According to the parameter change rule under fault, the system can build a fault database for the digital twin model, so as to speed up online diagnosis and fault prediction and provide guidance for the safe and stable operation of the marine steam power system.

To sum up, employing the digital twin technology, this paper uses modular modeling and multi-domain modeling methods to carry out digital modeling and simulation analysis of the marine steam power system. Based on the results of the test verification, steadystate characteristic verification, dynamic characteristic analysis and fault simulation, the dynamic characteristic analysis and state evaluation have been carried out, and the virtual reality mapping between the digital model and the physical equipment has been realized. This lays the foundation for mastering the dynamic characteristics of the marine steam power system.

Author Contributions

Writing-original draft, Methodology, Software, Visualization, G.Z.; Data curation, Formal analysis, J.W.; Writing-review & editing, Supervision, Funding acquisition, L.Z.; Formal analysis, Validation, X.X.; Investigation, Data curation, X.W., Writing-review & editing, Project administration, G.C. All authors have read and agreed to the published version of the manuscript.

Acknowledgments

Thanks to Guobing Chen (G.C.) and Lei Zhang (L.Z.) for their help with this article.

REFERENCES

1. Zhang, L.; Yang, Z.-C.; Liu, H.-R.; Chen, G.-B. Study on the characteristics of marine nuclear steam turbine units under coupling off design conditions and its influencing factors. Turbine Technol. 2019, 61, 266-270.
2. Nirbito, W.; Budiyanto, M.A.; Muliadi, R. Performance Analysis of Combined Cycle with Air Breathing Derivative Gas Turbine, Heat Recovery Steam Generator, and Steam Turbine as LNG Tanker Main Engine Propulsion System. J. Mar. Sci. Eng. 2020, 8, 726.
3. Perabo, F.; Park, D.; Zadeh, M.K.; Smogeli, Ø.; Jamt, L. Digital twin modelling of ship power and propulsion systems: Application of the open simulation platform. In

Proceedings of the 2020 IEEE 29th International Symposium on Industrial Electronics, Delft, The Netherlands, 17-19 June 2020; pp. 1265-1270.

4. Mrzljak, V.; Poljak, I.; Mrakovčić, T. Energy and exergy analysis of the turbo-generators and steam turbine for the main feed water pump drive on LNG carrier. Energy Convers. Manag. 2017, 140, 307-323.

5. Dulau, M.; Bica, D. Mathematical modelling and simulation of the behaviour of the steam turbine. Procedia Technol. 2014, 12, 723-729.

6. Chetan, M.; Yao, S.; Griffith, D.T. Multi-fidelity digital twin structural model for a sub-scale downwind wind turbine rotor blade. Wind Energy 2021, 24, 1368-1387.

7. Hu, M.-Y.; Kong, F.-L.; Yu, D.-L.; Yang, J. Digital Twin's Key Technologies and Application Prospects in the Field of Advanced Nuclear Energy. Power Syst. Technol. 2021, 45, 2514-2522.

8. Magargle, R.; Johnson, L.; Mandloi, P.; Davoudabadi, P.; Kesarkar, O.; Krishnaswamy, S.; Batteh, J.; Pitchaikani, A. A simulationbased digital twin for model-driven health monitoring and predictive maintenance of an automotive braking system. In Proceedings of the 12th International Modelica Conference, Prague, Czech Republic, 15-17 May 2017; pp. 35-46.

9. Fotias, N.; Bao, R.; Niu, H.; Tiller, M.; McGahan, P.; Ingleby, A. A Modelica Library for Modelling of Electrified Powertrain Digital Twins. In Proceedings of the 14th Modelica Conference 2021, Linköping, Sweden, 20-24 September 2021; pp. 249-261.

10. Zhou, L.; Mao, Z.-J.; Chen, Y.-M.; He, H.G. Digital Twin Construction Method of Satellite Communication Equipment Based on Multi domain Joint Modeling. In Proceedings of the China System Simulation and Virtual Reality Technology High Level Forum, Beijing, China, 15 December 2020; pp. 20-24.

11. Vering, C.; Mehrfeld, P.; Nürenberg, M.; Coakley, D.; Lauster, M.; Müller, D. Unlocking potentials of building energy systems' operational efficiency: Application of digital twin design for HVAC systems. In Proceedings of the Building Simulation 2019 Conference, Rome, Italy, 2-4 September 2019; pp. 1304-1310.

12. Xing, T.; Sun, L.F.; Wang, W.; Luo, W.C. Dynamics and Control Simulation Modeling and Flight Control Application of Digital Space Station. Aerosp. Control. Appl. 2021, 47, 40-47.

13. Mrzljak, V.; Poljak, I.; Prpić-Oršić, J. Exergy analysis of the main propulsion steam turbine from marine propulsion plant. Brodogr. Teor. Praksa Brodogr. Pomor. Teh. 2019, 70, 59-77.

14. Feng, H.; Tang, W.; Chen, L.; Shi, J.; Wu, Z. Multi-objective constructal optimization for marine condensers. Energies 2021, 14, 5545.

15. Zeng, G.-Q.; Wu, W.-H.; Chen, G.-B.; Li, J.; Wang, X.-F. Analysis of dynamic characteristics of marine deaerator. Energy Rep. 2022, 8, 121-129.

16. Yang, Y.-L.; Wu, W.; Wu, J.-X.; Zheng, Z.X. Simulation analysis on performance of fast variable load for marine steam power system. Chin. J. Ship Res. 2018, 13 (Supp. 1), 121-125.

17. Chang, Q.-Q.; Fu, X.-W.; Ding, H.-W.; Zhang, X.N. Study on the effect of low pressure supplementary steam on power distribution of marine double-cylinder steam turbine. J. Eng. Therm. Energy Power 2020, 35, 78-82.

18. Zhang, L.; Cao, Y.-Y.; Weng, L.; Cui, G.L. Dynamic characteristic analysis and fault diagnosis of low vacuum of marine condenser. Ship Ocean. Eng. 2017, 46, 67-71.

CHAPTER 3

Preliminary Study on Steam-Water Circulation System of Marine Small Nuclear Power Plant

ZHOU Long-yu, Hou Ying-zhe and Wu Hao

Wuhan Second Ship Design and Research Institute, Wuhan, Hubei,430064, China

ABSTRACT

Marine small nuclear power plants need to pursue the highest economy by improving the thermal efficiency of the steam-water circulation system. Based on the design concept of the steam-water circulation system of foreign nuclear-powered ships and onshore nuclear power plants, this paper proposes a preliminary heat balance diagram of the steam-water circulation system of marine small nuclear power plants, and comparatively analyzes the effect of initial steam pressure, condenser pressure and steam extraction pressure on thermal efficiency of steam-water circulation system. This study has certain guiding significance for the design of the general scheme of conventional islands in small marine nuclear power plants.

1. INTRODUCTION

Marine small nuclear power plants can independently supply power, desalinated water and heating for offshore facilities, such as comprehensive oil extraction supply stations and offshore oil drilling platforms.

Currently, most nuclear power ship steam-water circulation systems in China adopt the simplest heat balance process, so the thermal efficiency of conventional islands is low. Further, the marine small nuclear power plant uses technologies currently available in China. Therefore, we should learn the design concepts of nuclear power ships and nuclear power plant steam-water circulation systems from domestic or foreign research, which can optimize system composition and parameters and improve thermal efficiency.

Table 1. *Design parameters of foreign nuclear power ships.*

ship name	Savanna	Northern Line	Otto Han
country/ship type	American/passenger and cargo ship	Russian/nuclear-powered icebreaker	Germany/nuclear-powered merchant ships
reactor rated thermal power/MW	69	135	38
nuclear power plant efficiency/%	23.6	21.79	21.3
the steam parameters /MPa	2.8（Saturated Vapor）	3.9MPa/290°C	2.7MPa/268°C
condenser pressure /kPa	4.9	7.6	——
System Configuration	the three-level feed water heater（low-pressure heater, deaerator, high-pressure heater）	the four-level feed water heater（deaerator）	the three-level feed water heater
feed water temperature /°C	175	150~170	185

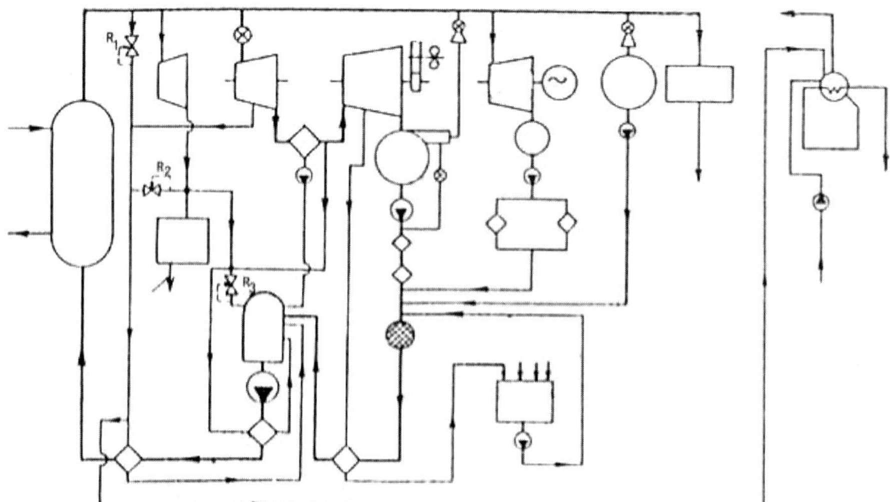

Figure 1. *The thermal balance diagrams of the Russian "Northern Line" nuclear-powered icebreaker*

2. FOREIGN NUCLEAR POWER SHIPS' THERMAL BALANCE DIAGRAM OF THE STEAM-WATER CIRCULATION SYSTEM

In terms of the design of the heat balance diagram of the steam-water circulation system of the marine small nuclear power plant, we can refer to the steam-water circulation system of nuclear power ships and onshore nuclear power plants in other countries. For example, the thermal balance diagrams of the Russian "Northern Line" nuclear-powered icebreaker, the American "Savanna" and the Germany "Otto Han" nuclear-powered merchant ships are shown in Figures 1,2 and 3[1,2]. The design parameters of their conventional steam-water circulation system are shown in the table below.

The thermal balance process of the steam-water circulation system of the "Northern Line" nuclear-powered icebreaker has three steps. First, the steam parameters at the outlet of the steam generator are 3.92MPa and 290°C. When the steam reaches the steam turbine, the steam parameters become 3.6 MPa, 280°C. Second, the steam enters the high-pressure cylinder. After passing through the high-pressure cylinder, the steam enters the intermediate steam-water separator. Third,

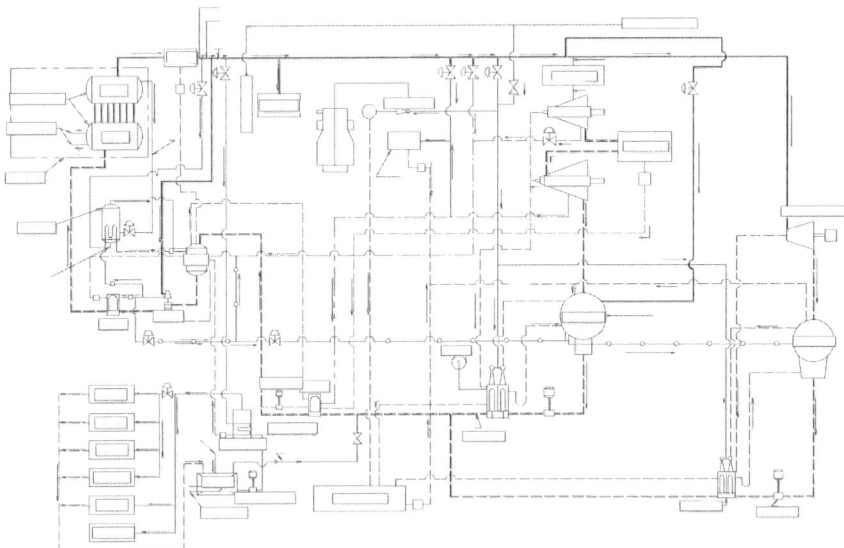

Figure 2. *The thermal balance diagrams of the American "Savanna" passenger and cargo ship*

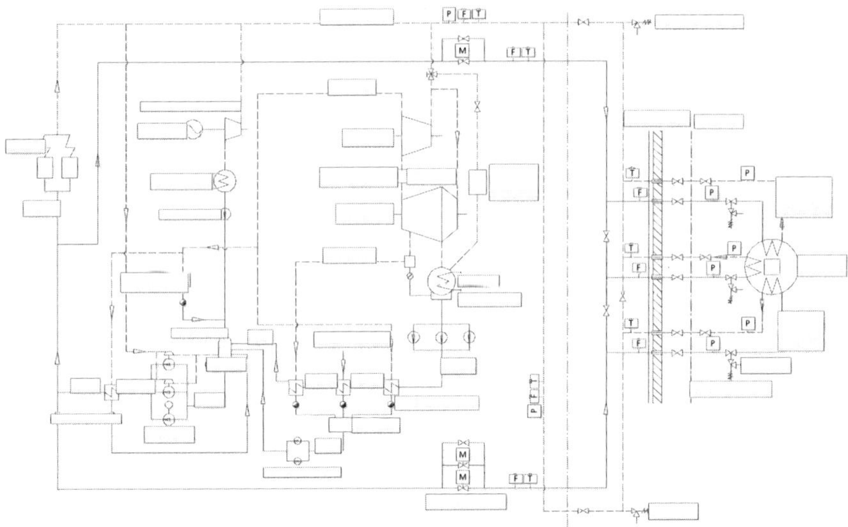

Figure 3. *The thermal balance diagrams of the Germany "Otto Han" nuclear-powered merchant ships*

the steam enters the low-pressure cylinder, and the residual steam of the steam turbine is pumped to the deaerator, distillation device and the first-stage feed water heater. In doing so, the excess steam is discharged to the main condenser and cooling condenser where the pressure is maintained at 0.0076MPa. Based on the main condensate electric pump, the condensate is pumped to the deaerator after passing through the main ejector, suction ejector, ion exchange filter, and low-pressure feed water heater. The feed water passing through the deaerator has a pressure of 0.13MPa and a temperature of 106°C. The feed water that passes through the high-pressure feed water heater has a temperature of 150°C ~ 170°C. The feed water is sent to the steam generator through the main feed water vapour pump.

The backup feed water pump installed can provide feed water of 3.5MPa from the deaerator to the steam generator. When the device operates or stops, the steam and steam-water mixture generated by the steam generator is received by two auxiliary condensation devices.

The three-stage feed water heating is performed by the steam extraction of the steam turbine unit. (The three-level feed water heater is: low-pressure heater, deaerator, high-pressure heater): the first-level feed water heater (low-pressure heater) is heated by steam with a pressure of 0.083MPa extracted from the extraction branch of the low-pressure cylinder, and then the generated condensate water is discharged to the hot condensate tank and drained to the main condensation pipe by drain pump.

The steam pressure of the steam branch of the steam-water separator reaches 0.32MPa. After passing through the intermediate steam-water separator, the steam and the residual steam generated by the steam turbine feed pump will be supplied to the deaerator, the second-level feed water heater and the distillation device.

The third-level feed water heater uses the extraction steam from the extraction branch of the high-pressure cylinder. The steam extraction pressure is 0.98MPa. After condensation in the heater, the steam will be discharged to the deaerator.

The nuclear-powered merchant ship "Savanna" and the nuclear-powered ore ship "Otto Han" basically share the same thermal balance diagram of the steam-water circulation system as that of the "Northern Line". The difference is that "Savanna" and "Otto Han" are equipped with two instead of three extraction points, and a three-stage feed water heater (the low-pressure heater of "Otto Han" contains a two-stage hydrophobic regenerator [3]).

The above-mentioned heat balance diagrams share two features in common:

(1) 2~3 steam extraction points and 3~4-level feed water heaters are set on the main steam turbine unit, and the latent heat of vaporization of the steam turbine unit is fully utilized to improve the thermal efficiency of the power plant;
(2) An emergency feed water pump is equipped to meet the water demand of the steam generator in case of failure of the steam turbine feed water pump.

The land-based PWR nuclear power plant is a high-power reactor, and its construction space is not limited, so its initial vapour pressure is large (generally 5~6 MPa), and it will be set at 5 7-level heaters. The biggest difference with nuclear-powered ships is that the onshore pressurized water reactor nuclear power plant is equipped with a steam-water separation reheater (MSR), which is absent in the steam-water circulation system of nuclear-powered ships . Figure 4 is the heat balance diagram of a 900MW pressurized water reactor nuclear power plant. The initial steam pressure is 5.47MPa, and MSR is adopted, and a 5-level heater is installed, and the gross efficiency of the power plant is 33.14%.

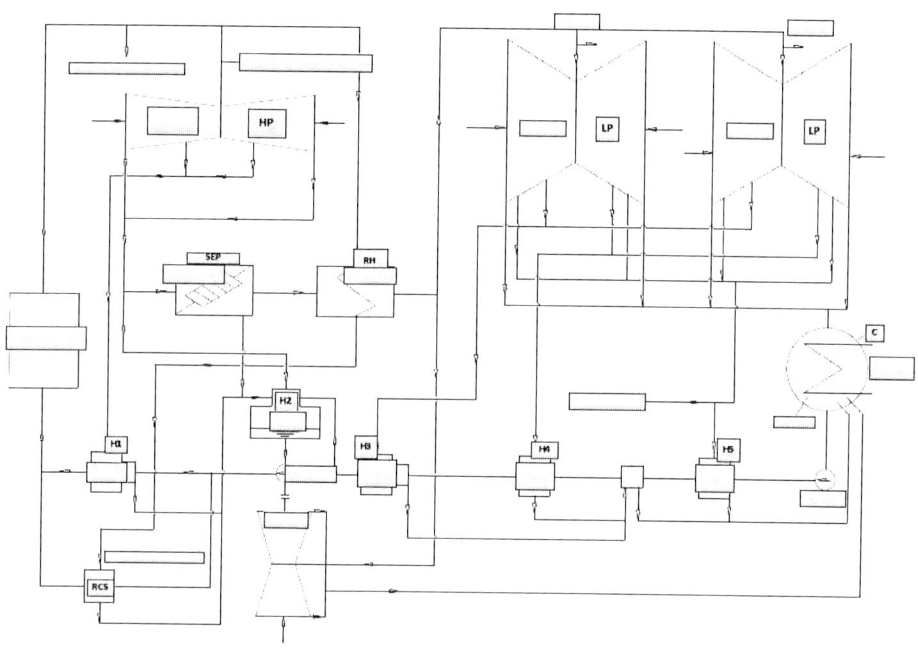

Figure 4. *The thermal balance diagram of 900MW PWR nuclear power plant unit*

Table 2. *Steam generator performance comparison*

	OTSG	SG
advantage	Compact structure High thermal efficiency Fast load response	Mature technology Good inheritance
disadvantage	High requirements for feed water quality indicators High requirements on nuclear control system Need to set start and stop system	Large size Low thermal efficiency Slow load response

Actually, increasing the initial steam pressure is beneficial to improving the thermal efficiency of the steam-water circulation system, so some researchers have proposed the use of a once-through steam generator. Table 2 shows the performance comparison between the once-through steam generator (OTSG) and the natural circulation steam generator (SG).

The over-current steam generator has not been used in domestic nuclear facilities, and thus no experience in this regard is available in China. Therefore, the reliability and safety of the operation of the once-through steam generator should be explored. The control of water supply coordination, system start and stop and discharge, system monitoring and other aspects will be different if a direct-current

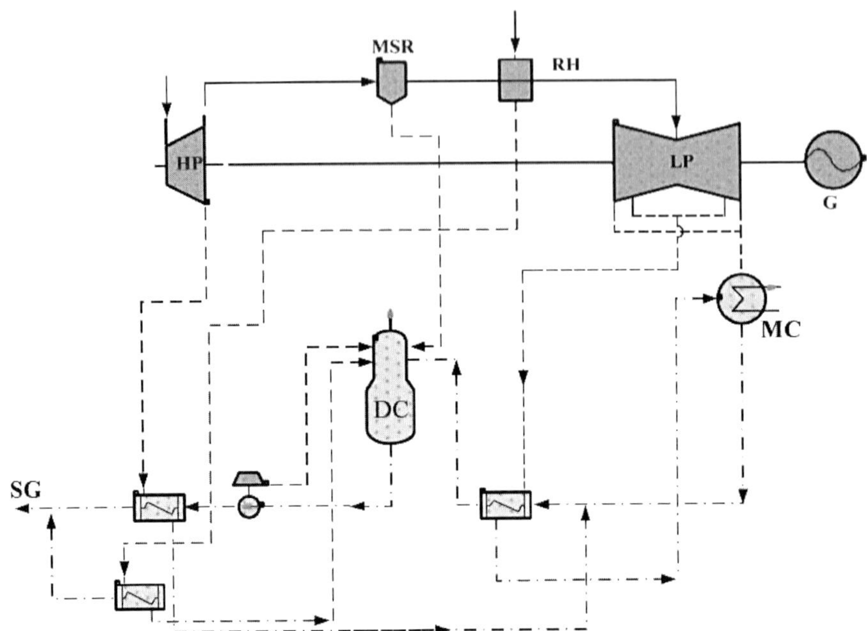

Figure 5. *The thermal balance diagrams of the steam-water circulation system of marine small nuclear power plant*

steam generator is used. Thereby, how to achieve safe and reliable control is one of the key issues in the design of steam water circulation systems.

Based on the above analyses, it is advisable to employ natural circulation steam generators, and add MSR based on the existing mature technology, and set 2~3 steam extraction points, and 3~4-level heaters to improve thermal efficiency of steam-water circulation systems in marine small nuclear power plants.

3. ANALYSIS OF SOME FACTORS INFLUENCING THE THERMAL EFFICIENCY OF THE STEAM-WATER CIRCULATION SYSTEM OF A SMALL MARINE NUCLEAR POWER PLANT

According to the mature design experience of nuclear power ships and the partial design of onshore nuclear power plants abroad, it is suggested that China should consider the performance requirements and technical characteristics of small marine nuclear power plants. They should initially utilize a natural steam generator, add MSR, adopt Level 2 steam extraction recuperation and the Level 3 feed water heating scheme.

3.1. The effect of initial steam pressure on the thermal efficiency of the steam-water circulation system

The steam parameters at the outlet of the steam generator have a significant effect on the thermal efficiency of the steam-water circulation system. This paper

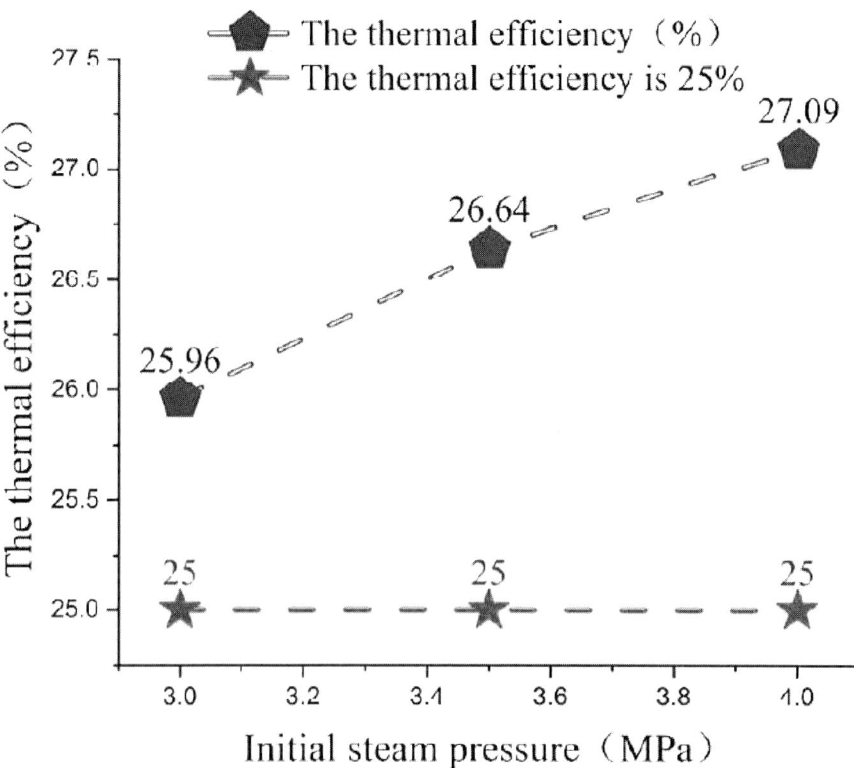

Figure 6. *Relationship between initial steam pressure and thermal efficiency of the steam-water circulation system*

calculates the thermal efficiency of the steam water circulation system when the initial steam pressure is 3.0MPa, 3.5MPa and 4.0 MPa.

As Figure 6 shows, when initial steam pressure is within a certain range, the thermal efficiency of steam-water circulation system increases with the increase of initial steam pressure and the maximum efficiency is 27.09%. However, the following problems will be incurred:

(1) The steam humidity at the steam turbine outlet would increases. If the steam turbine exhaust humidity rises excessively, erosion will occur on the last few blades of the steam turbine, which not only adversely affects the operation safety of the steam turbine, but also reduces the internal efficiency of the steam turbine;
(2) Increasing the operating pressure of the steam generator will require an increase in the head of the feed water pump. As a result, the axial thrust of the impeller of the feed water pump rises, and the power consumption of the feed water pump also increases with regard to the unit mass flow of working fluid;
(3) Increasing the initial steam pressure is also subject to the influence of the operating temperature of the coolant on the primary side of the steam generator.

Table 3. *Calculation result of the effect of condenser pressure on the thermal efficiency of steam-water circulation system*

condition / parameter	1	2	3
SG outlet pressure (MPa)	3.5	3.5	3.5
heat exchange area (m^2)	1805.72	1490.53	1274.84
condenser pressure (kPa)	5.5	6.0	6.5
the thermal efficiency (%)	26.81	26.64	26.50

3.2. The effect of condenser pressure on the thermal efficiency of the steam-water circulation system

According to the Carnot cycle thermal efficiency formula, reducing the condenser pressure can promote the thermal efficiency of the steam-water circulation system more effectively. In this study, the steam-water circulation thermal efficiency is calculated when the initial steam pressure is 3.5MPa and the condenser pressure reaches 5.5kPa, 6.0kPa and 6.5kPa, respectively. The calculation results are shown in Table 3.

The calculation results show that when the condenser pressure is 5.5kPa, the thermal efficiency of the steam-water circulation system is the highest, reaching 26.81%, but the heat exchange area of the condenser increases sharply: the heat exchange area in this scenario is 29.4% higher than when the condenser pressure is 6.5kPa, which has a greater impact on the overall layout and economy of the system.

3.3. The effect of steam extraction pressure on the thermal efficiency of the steam-water circulation system

The steam-water circulation system recycles the latent heat of vaporization of residual steam by providing steam extraction in a high-pressure cylinder. If the steam extraction parameters and the steam extraction position are different, then it also has different impacts on the thermal efficiency of the steam-water circulation system. In terms of the problem, when the initial steam pressure is 3.5MPa, and the condenser pressure is 6.0kPa, this paper calculates different extraction schemes with extraction points set at different positions in the high-pressure cylinder stage. The calculation results are shown in Table 4.

The calculation results show that when the heat transfer end difference of the feed water heater is controlled at 8 ~ 15°C, the thermal efficiency of the steam-water circulation system is high. When the pressure value is not in this range, the higher the steam extraction pressure of the high-pressure cylinder, and this is not conducive to improving the thermal efficiency of the steam-water circulation system. Therefore, the optimal steam extraction pressure of the steam-water circulation system is closely related to the feed water temperature of the steam generator.

Table 4. *Calculation result of the effect of steam extraction pressure on the thermal efficiency of steam-water circulation system*

parameter \ condition	1	2	3
SG outlet pressure（MPa）	3.5	3.5	3.5
Steam extraction pressure in high-pressure cylinder stage（MPa）	0.60	0.76	0.98
Saturation temperature corresponding to Steam extraction pressure（°C）	158.83	168.30	179.01
the thermal efficiency（%）	26.64	26.63	26.54

4. CONCLUSION

The paper researched the design methods of the steam-water circulation system of nuclear-powered ships and nuclear power plants at home and abroad. In addition, we proposed a preliminary heat balance diagram of the steam-water circulation system of marine small-scale nuclear power plants. The paper also analyzed some factors affecting the thermal efficiency of the steam-water circulation system, and the following conclusions are as follows.

By studying the design methods of the steam-water circulation system of nuclear-powered ships and nuclear power plants at home and abroad, this paper proposes a preliminary heat balance diagram of the steam-water circulation system of marine small nuclear power plants and analyzes some factors that affect the thermal efficiency of the steam-water circulation system. Finally, the article concludes some results as follows.

1) The marine small nuclear power plants not only focus on the economy but also are restricted by its weight and size. Therefore, it is recommended to use natural circulation steam generators for the steam-water circulation system and to install MSR dependent upon the basis of mature technology. We still need to set 2~3 steam extraction points as well as 3~4-level heaters.
2) The thermal efficiency of the steam-water circulation system increases with the increase of initial steam pressure, but it will also result in some problems and be limited by steam initial pressure value provided by the nuclear island.
3) The pressure of the condenser has a great impact on the thermal efficiency of the steam-water circulation system. Although reducing its pressure can improve the thermal efficiency, the heat exchange area of the condenser also increases, which has a greater impact on the overall layout and economy.
4) The optimal steam extraction pressure is closely related to the steam generator feed water temperature, and it is of benefit to the later design of the steam turbine unit.

REFERENCES

1. Safety analysis report of the North Line Atomic Icebreaker(1995).
2. Peng M.J. (2009) Ship nuclear power plant. Atomic Energy Press, Beijing.
3. Zang X.N. (2010) Nuclear Power Plant Systems and Equipment. TSINGHUA UNIVERSITY PRESS, Beijing.

Prediction of External Exposure during Dismantling of Steam Generator

Martin Hornáček and Vladimír Nečas

Institute of Nuclear and Physical Engineering, Slovak University of Technology in Bratislava, Slovakia

ABSTRACT

The decommissioning of nuclear power plants is in the Slovak Republic an actual issue. In 2015 started the second decommissioning stage of nuclear power plant V1 in Jaslovské Bohunice. This stage involves the dismantling and segmentation of activated (reactor pressure vessel, reactor internals) and contaminated parts (steam generators, pressurizer). From this reason it is necessary to investigate the radiation situation in the vicinity of the component to be cut. The presented results show that during remote dismantling the exposure is small (compared with the fragmentation tasks). Moreover, when the pre-dismantling decontamination with decontamination factor of 100 is applied, the total collective effective dose is below the yearly limit of 20mSv *for workers.*

Keywords: *Steam Generator; Dismantling and FragmentationDismantling and Fragmentation; External ExposureExternal Exposure*

1. INTRODUCTION

During decommissioning of nuclear power plants (NPP) several steps have to be carried out in order to achieve the desired end state in the planned time schedule. This includes mainly preparation for dismantling, dismantling of technological equipment, treatment and conditioning of resulting waste, storage and the final disposal of radioactive waste and in many cases also the decontamination. It is obvious that during these activities, the proper knowledge of the radiation situation and thus the optimization of radiation protection within the ALARA principle is crucial (this principle means that the exposure, the number of exposed persons and the probability of exposure should be "As Low As Reasonably Achievable").

In case of so-called large components (e.g. activated parts-reactor pressure vessel and internals; contaminated parts-pressurizer, steam generator-SG) the usual dismantling and transportation techniques have to be modified and many factors like site-specific conditions, availability of technical tools, the strategy of radioactive waste disposal have to be taken into account [1].

In the Slovak Republic, the NPP V1 in Jaslovské Bohunice is currently in the 2nd and final decommissioning stage with planned duration between 2015-2025 [2]. In this NPP, the VVER-440/230 reactor type (Russian type of pressurized water reactor) was used. Each unit had gross electrical output of 440MW and the cessation of operation was after 28 years of standard operation (1978-2006 and 1980-2008).

Within the second decommissioning stage, the large components will be cut in-situ and the fragmented parts will be stored and/or conditioned and disposed in the repository [3].

2. TECHNICAL DESCRIPTION OF STEAM GENERATOR USED IN NPPS WITH VVER-440 TYPE REACTOR

The subject of the analysis in this paper is the steam generator used in each of the 6 loops of the primary circuit within one unit-Figure 1.

From the construction point of view the vessel is made of carbon steel 22 K; the collector material as well as the heat exchanging tube material is titanium stabilized austenitic steel with 0.08% carbon, 18% chromium, 10% nickel and less than 1% titanium [5].

3. CALCULATION TOOL VISIPLAN 3D ALARA

The calculation of external exposure during dismantling and fragmentation of the SG was realized using code VISIPLAN 3D ALARA developed by Belgian company SCK-CEN. The photon fluency rate at the dose point near the volume source can be determined by considering the volume source as consisting of a number of point sources. By adding the contribution of every point source to the dose at the dose point the photon fluency rate at the dose point is expressed as [6]:

$$\varphi = \int_V \frac{S \cdot B \cdot \mathrm{e}^{-b}}{4\pi\rho^2}\, \mathrm{d}V \left(\mathrm{m}^{-2} \cdot \mathrm{s}^{-1}\right) \tag{1}$$

where: S-source strength per unit volume $\left(\mathrm{n} \cdot \mathrm{s}^{-1}\right)$,

ρ-distance from a point source (m),

B-buildup coefficient (-),

b-attenuation effectiveness coefficient (-),

V-volume $\left(\mathrm{m}^3\right)$.

Each small source is called a kernel and the process of integration, where the contribution to the dose of each point is added up, is called "point kernel" integration. This is the method used in the VISIPLAN software [6]. Based on the photon fluency rate at a point it is possible to calculate the effective dose rate depending on the dose conversion factors selected in the calculations [6]:

$$E = \sum_i h_i \cdot \phi_i(\mathrm{Sv/s}) \tag{2}$$

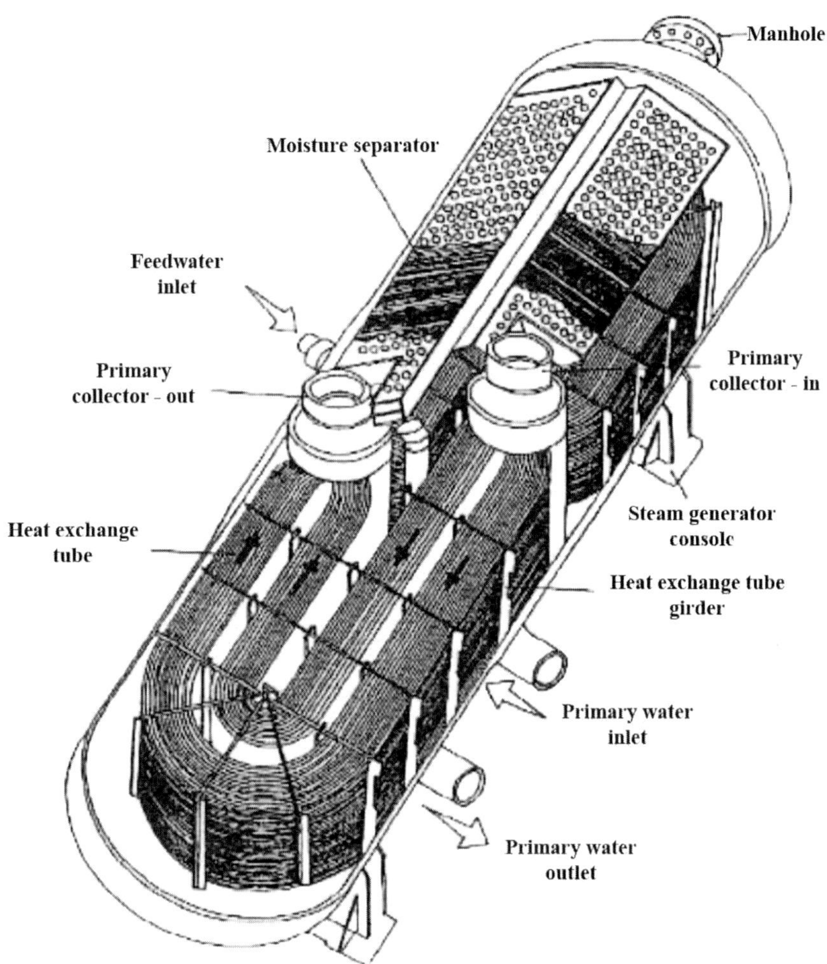

Figure 1. *Steam generator for NPPs with VVER-440 type reactor [4].*

Table 1. *Considered activity values for parts of steam generator.*

Component	Activity of γ and X-ray emitting nuclides [Bq]		
	2015	2020	2025
Heat exchange tubes	19.2E+10	10.7E+10	6.32E+10
1 Collector	18.6E+07	10.4E+07	6.15E+07

where: h_i —dose conversion coefficient for photons of energy (Sv per photon/m^2),
ϕ_i-fluency rate of the photons at energy $\left(\mathrm{m}^{-2} \cdot \mathrm{s}^{-1}\right)$.

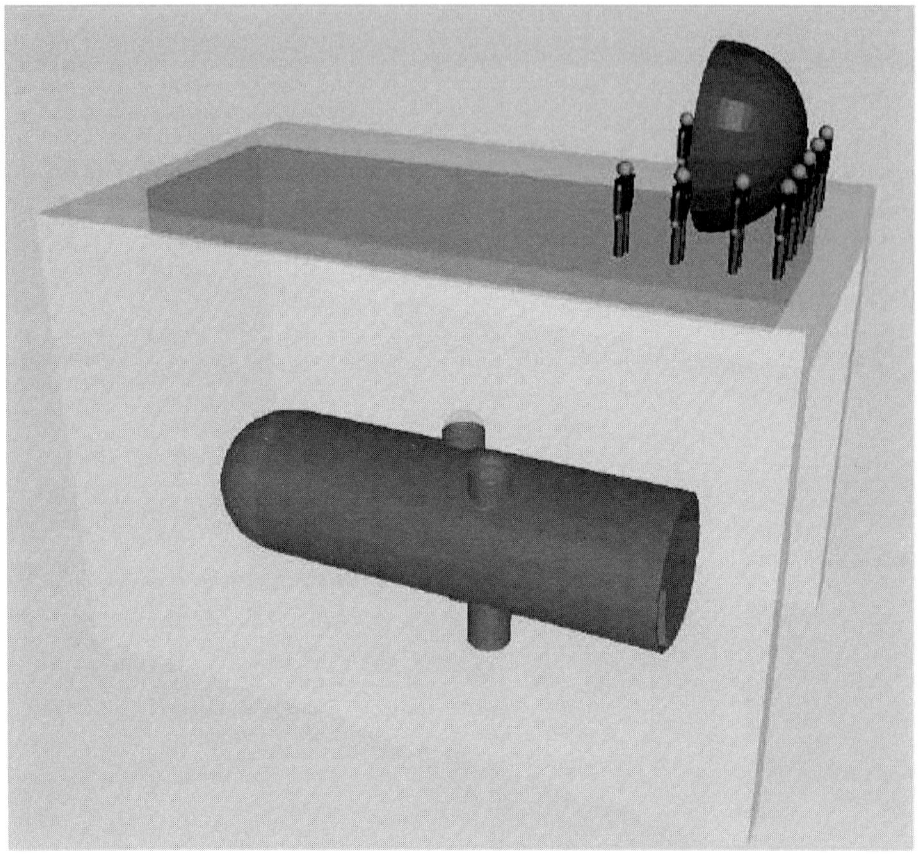

Figure 2. *Cutting the end part of steam generator-top view.*

4. GENERAL PRECONDITIONS AND INPUT DATA

The chapter deals with the model of steam generator, considered input data for the calculations and considered worker group.

4.1. Model of Steam Generator

The considered model of steam generator consists of the following parts:

- SG casing - consisting of cylindrical part (length approx. 9.5 m) and two semi-spherical parts (radius approx. 1.7 m), total mass 113.4t, part of secondary circuit—negligible level of activity.
- Heat exchange tubes (5536 pieces of U-tubes—modelled in simplified geometry), the length approx. 9.7 m, total mass approx. 34.7t, part of primary circuit-contaminated parts.
- 2 collectors - the height of each 4.2 m, outer diameter 0.97 m, the mass of 12.7t each, part of primary circuit-contaminated parts.

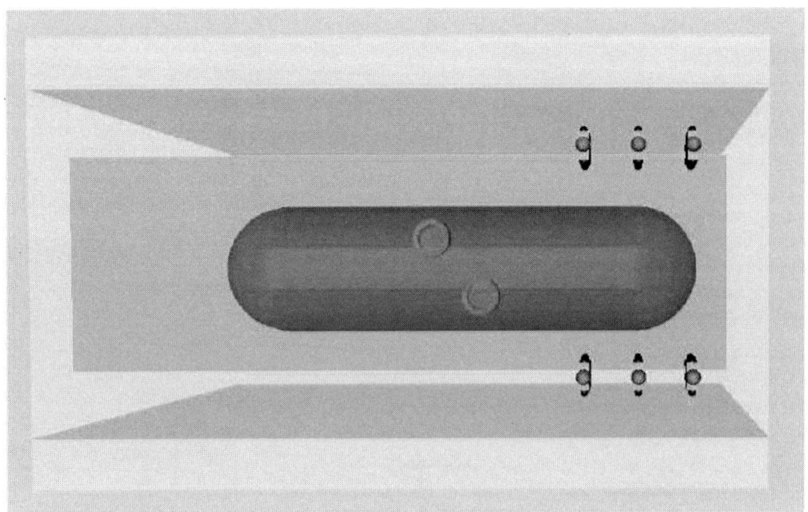

Figure 3. *Lifting the fragmented part.*

Table 2. *Remote dismantling of steam generator-total in 2015.*

Task	Workload [manh]	Average dose rate [mSv/h]	Collective effective dose [manmSv]
Cutting the end part of steam generator	35	2.52E−03	8.80E−02
Lifting the fragmented part	1	4.21E−03	4.21E−03
Cutting the segment of SG casing and heat exchange tubes	160	3.57E−03	5.72E−01
Lifting the fragmented segment	7.5	1.34E−01	1.00E+00
Tilting the collectors to the horizontal position	0.5	5.73E−05	2.87E−05
Lifting the collectors	1.1	3.94E−04	4.34E−04
Total	**205.1**	-	**1.7**

Table 3. *Fragmentation of the segments-total in 2015.*

Task	Workload [manh]	Average dose rate [mSv/h]	Collective effective dose [manmSv]
Fragmentation of casing of the segment	300	2.99E−01	9.04E+01
Fragmentation of heat exchange tubes of the segment	500	8.84E−01	2.23E+02
Fragmentation of collectors	320	1.68E−03	5.41E−01
Total	**1120**	-	**313.9**

Table 4. *The inluence of time-general comparison.*

Task	Total collective effective dose [manmSv]		
	2015	**2020**	**2025**
Remote dismantling of steam generator	1.7	0.9	0.5
Fragmentation of the segments	313.9	164.9	88.4
Total	**315.6**	**165.8**	**88.9**

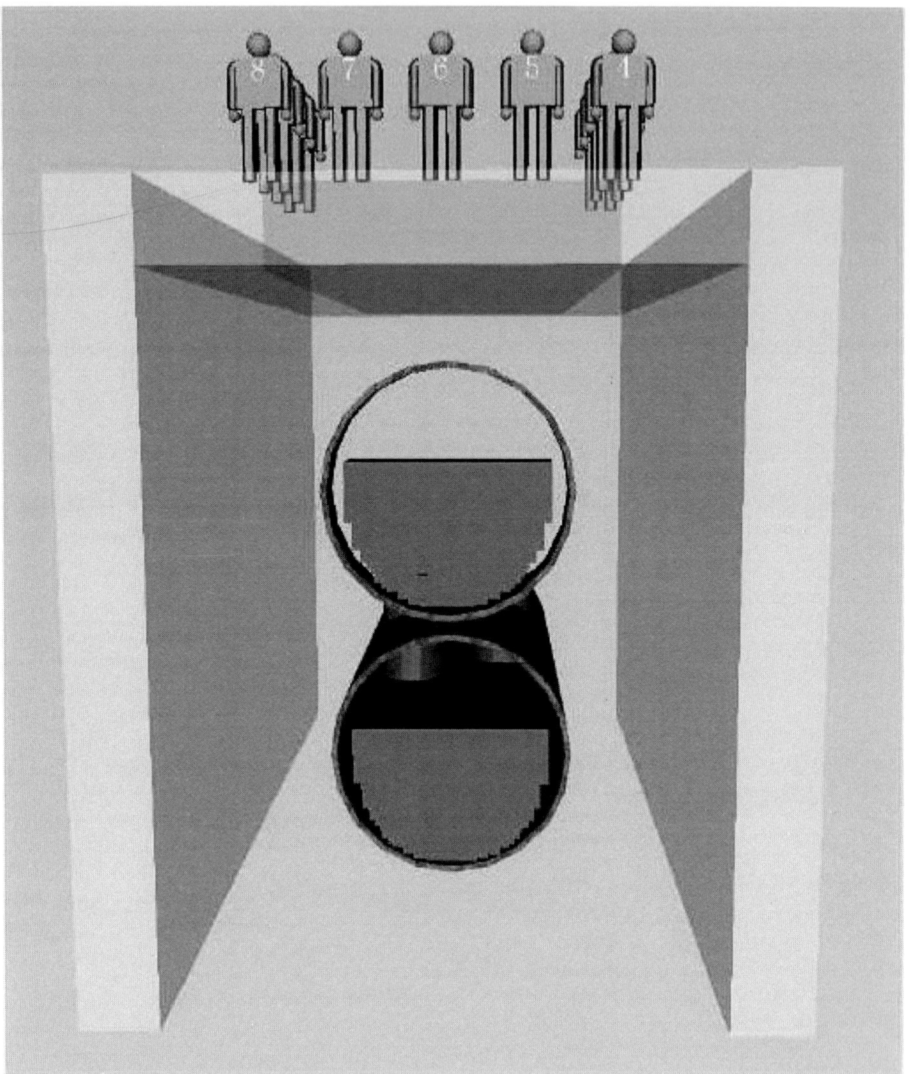

Figure 4. *Lifting the fragmented segment—side view.*

The total mass of SG is then approx. 173t.

The steam generator is placed in so-called hermetic box made of concrete walls with thickness of 1 m. The distance between the reactor floor and SG's axis is 7.4 m; the distance between casing and walls is 1.7 m.

4.2. The Source Term

Among the geometric dimensions and material composition the other crucial input parameter is the source term, i.e. the level of activity of each part and the

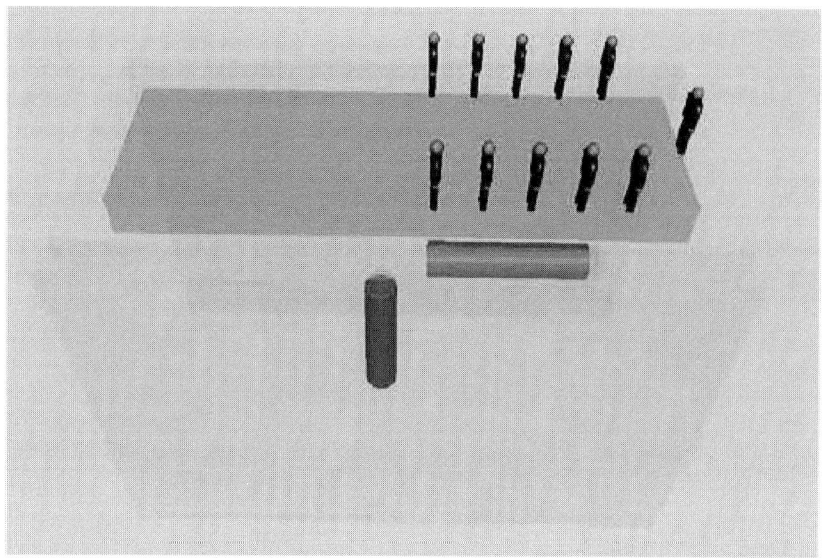

Figure 5. *Lifting the collector.*

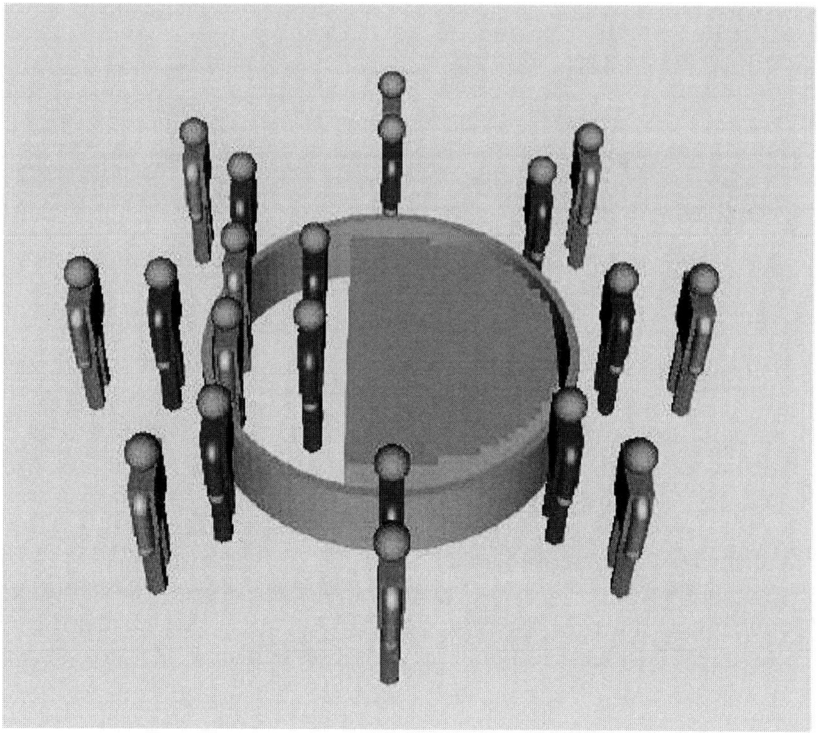

Figure 6. *Fragmentation of casing of segment.*

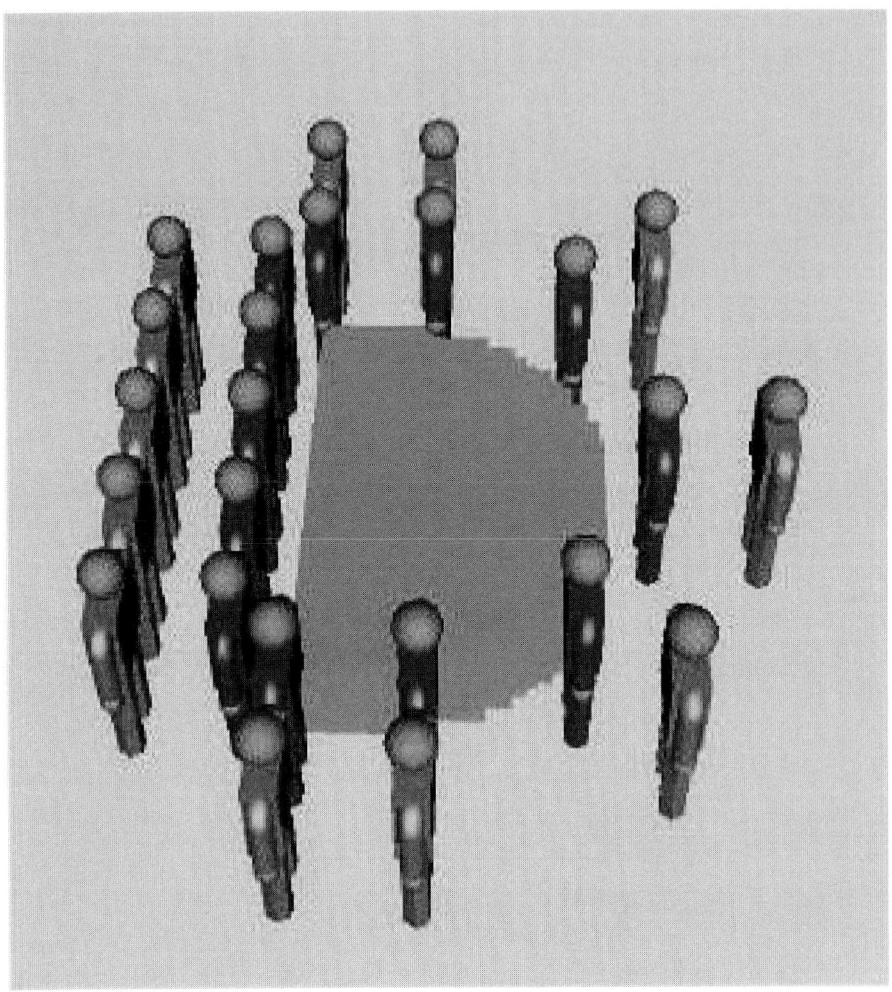

Figure 7. *Fragmentation of heat exchange tubes.*

nuclide composition of the source (nuclide vector). The activity content of each contaminated part of SG represents estimated values for calculation of decommissioning parameters and is depicted in Table 1.

It is necessary to emphasize that the activity values depicted in Table 1 are without applying decontamination.

The nuclide vector is derived from the radiological characterization and the most dominant nuclide is ^{60}Co with approx. 90% share (when considering γ and X-ray emitting nuclides)-result of standard operation with no leakages of the primary circuit.

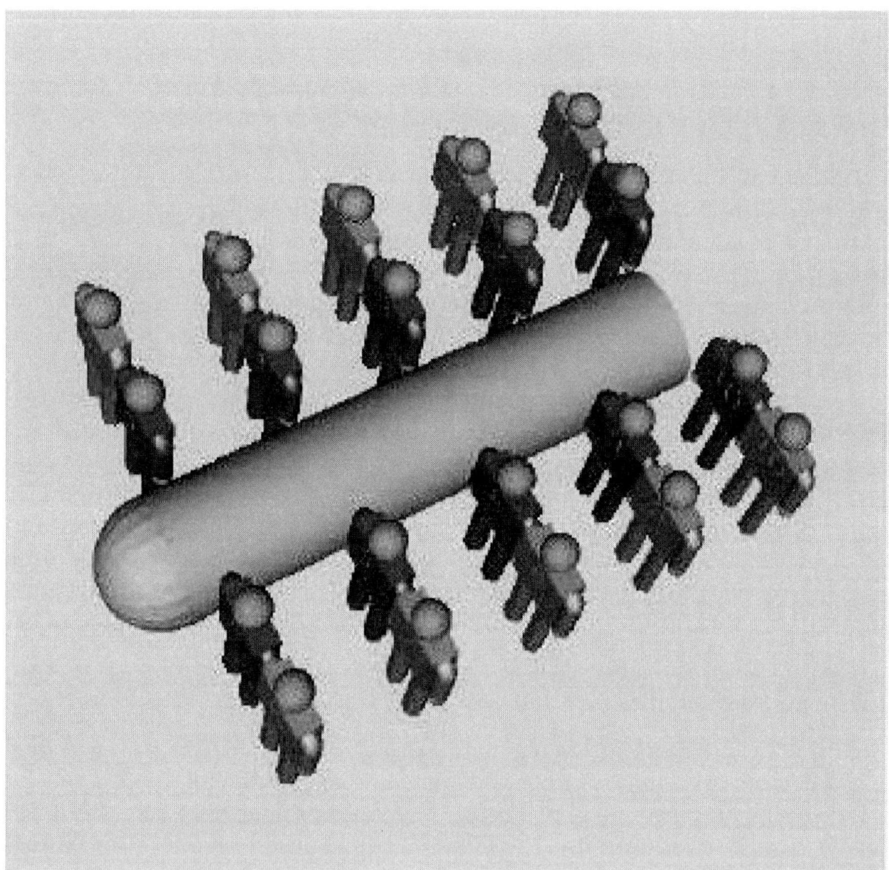

Figure 8. *Fragmentation of collector.*

4.3. Worker Group

During the cutting of steam generator and lifting the segmented parts from the hermetic box to the reactor floor, only one trajectory (set of points where the dose rate is calculated) is considered. In case of fragmentation of the segmented parts, the following worker group is considered:

- Cutter and junior technician-realization of cutting activities - the distance from the component is 30 cm, the time coefficient (considering the time of stay during each activity) is 1.
- Master and technician-management of workers, control of exposure time (master), control of the cutting techniques, the quality and speed of the cut (technician)_the distance from the component is 100 cm, the time coefficient is 0.8 .

- Radiation protection technician-monitoring radiological situation, measurement of dose rate-the distance from the component is 100 cm, the time coefficient is 0.3 .

5. CONSIDERED DISMANTLING PROCEDURE

The dismantling of steam generator is considered to be realized in the hermetic box (remote dismantling with control on the reactor floor). The dismantled parts will be lifted to the reactor floor where their fragmentation will be carried out. The whole procedure is described below (in case of remote dismantling the blue person are depicted, in case of fragmentation activities the trajectory of cutter and junior technician is depicted by red persons, the other trajectories are depicted by the green persons).

In order to avoid aerosol production, the cutting using band saw is considered. The duration of each step was derived from the cutting speed of the facility for fragmentation of large components in the Interim Storage North at the Greifswald site, which varies from 15 to 80 mm/min [7]. Within the conservative approach and since the procedure is performed in the controlled area, the lowest speed from the interval was chosen. The issue of cutting duration was also discussed with professionals [8]. Based on these data (using the time coefficients described in Chapter 4.3 as well) the time of duration of each activity and also the time of stay of each worker can be obtained.

1. Dismantling activities

- Cutting the end part of steam generator-total workload 35 man-hours (Figure 2). The hole in the reactor floor is depicted as blue rectangle.
- Lifting the fragmented part (Figure 3)_total workload of 1 man-hour. This value was derived from the considered lifting speed of 5 mm/s and the distance from the reactor floor (approx. 9 m).
- Cutting the segment of SG casing and heat exchange tubes-similar situation as depicted in Figure 2. Based on the data from the available technical documentation, the average distance between girders of heat exchange tubes is 68.8 cm, thus the width of the fragments is considered to be 70 cm (in order to achieve the fixation of heat exchange tubes during and after cutting as well as during lifting). The total workload is 160 man-hours.
- Lifting the fragmented segment-total workload of 7.5 man-hours-Figure 4.
- Tilting the collector to the horizontal position-workload of 0.25 man-hours for each collector.
- Lifting the collector-workload 0.55 man-hours for each collector. The situation is depicted in Figure 5.

2. Fragmentation activities

- Fragmentation of casing of the segment-based on the fact that the casing of steam generator is considered to be released into the environment, this part is fragmented first (in order to remove the material and to enable enough space for later fragmentation of heat exchange tubes). The total workload for 15 segments is 300 man-hours, the situation is depicted in Figure 6.

- Fragmentation of heat exchange tubes of the segment-The total workload for 15 segments is 500 man-hours, the situation is depicted in Figure 7.
- Fragmentation of collectors-The total workload is 160 man-hours for each collector-Figure 8. 3) Fragmentation activities
- Fragmentation of casing of the segment-based on the fact that the casing of steam generator is considered to be released into the environment, this part is fragmented first (in order to remove the material and to enable enough space for later fragmentation of heat exchange tubes). The total workload for 15 segments is 300 man-hours, the situation is depicted in Figure 6.
- Fragmentation of heat exchange tubes of the segment-The total workload for 15 segments is 500 manhours, the situation is depicted in Figure 7.
- Fragmentation of collectors-The total workload is 160 man-hours for each collector-Figure 8.

6. RESULTS

The results of the calculations are presented in Tables 2-4. As was mentioned in the beginning of the paper, the 2nd decommissioning stage is planned between 2015 and 2025. From this reason, the influence of time on the exposure was studied as well_-Table 4. For lucidity, only total results are presented in Table 3 and Table 4. The issue of fragmentation of the segments was analyzed more in detail in [9].

7. CONCLUSION

From the results depicted in Tables 2 – 4 it can be stated that biggest contribution to the total collective effective dose is from the activities regarding the fragmentation of the segments. This can be expected because these tasks are carried out in the vicinity of the most contaminated part-heat exchange tubes.

The general overview in the Table 4 shows that the time period of 5 years results in approx. 1.9-times decrease of total collective effective dose. This is in accordance with the half-life of ^{60}Co (5.27 a) which is the most dominant nuclide within the considered nuclide vector.

The calculated dose rates and thus the doses are quite high. As was mentioned in Chapter 4.2, no pre-dismantling decontamination is considered. This is, however, a very conservative assumption. Several projects regarding decontamination of the primary circuit of shut-down NPPs with pressurized water reactors were realized, for instance, in German NPP Unterweser (1345 MWe, more than 30 years of operation) where the decontamination factor (DF) of 147 of SG tube section was achieved [10]. When considering the DF of 100, it can be expected that the dose rates will be 2 orders of magnitude lower. This will be probably also the case of NPP V1 in Jaslovské Bohunice, where the decontamination of the primary circuit is realized.

During the creation of calculation models and trajectories the site-specific conditions were taken into account which enables the application of the results in the planning of the dismantling process and fragmentation activities together with their optimization in accordance with the ALARA principle.

Acknowledgements

This project has been supported by the Slovak Grant Agency for Science through grant VEGA 1/0796/13.

REFERENCES

1. Organisation for Economic Co-Operation and Development-Nuclear Energy Agency. Radioactive Waste Management Committee (2012) The Management of Large Components from Decommissioning to Storage and Disposal: A report of the Task Group on Large Components of the NEA Working Party on Decommissioning and Dismantling (WPDD). NEA/RWM/R(2012)8. Paris: OECD/NEA.
2. National Nuclear Found for Decommissioning of Nuclear Installations and for Spent Fuel and Radioactive Waste Management (2012) Strategy of Back-end of Peaceful Use of Nuclear Energy in Slovak Republic. (In Slovak)
3. Nuclear and Decommissioning Company (2013) The Intent in terms of the Act No. 24/2006 Coll. on Environmental Impacts Assessment and Alternations and Amendments of certain Acts as Amended-The 2nd Stage of Decommissiong of V1 NPP Jaslovské Bohunice. (In Slovak) http://www.javys.sk/sk/o-spolocnosti/projekty/posudenie-vplyvu-2-etapy-vyradovania-je-v1-na-zp/dokumenty
4. Roupec, P. (2010) Transfer Heat Analysis in Steam Generators in Blocks of VVER 440. Master's thesis, Energy Institute, Faculty of Mechanical Engineering, Brno University of Technology. (In Czech)
5. International Atomic Energy Agency (1997) Assessment and Management of Ageing of Major Nuclear Power Plant Components Important to Safety: Steam generators. IAEA-TECDOC-981. ISSN 1011-4289, pp. 15-17.
6. Vermeersch, F. (2005) Dose Assessment and ALARA Calculation with VISIPLAN 3D ALARA Planning Tool, Training Course. IDPBW Nuclear Studies, Boeretang: SCK.CEN, Belgium.
7. Rohde, M. (2011) Treatment and Conditioning of Dismantled Material and Operation Waste in EWN, Overview. http://www.iaea.org/OurWork/ST/NE/NEFW/WTS-Networks/IDN/idnfiles/CuttingTechniqueWkp-Germany2011/Treatment-and-Conditioning-of-DismantledMaterial-and-OperationWaste.pdf
8. Personal and via E-Mail Consultation with Mr. Andreas Hammel from the Company WIKUS within the Seminar "Fachtagung Rückbau—Fachtagung zur Technik und Praxis des Rückbaus von Kernkraftwerken" organised by Kraftwerksschule e.v. in NPP Zwentendorf, Austria, from 24-25 September 2014.
9. Hornacek, M. and Necas, V. (2015) The Assessment of NPP V1 Steam Generator Dismantling from the Perspective of External Exposure. Proceedings of International Conference ECED 2015-Eastern and Central European Decommissioning, 23-25 June 2015, Trnava, Slovakia.
10. Topf, Ch., Belda, L.S., Fischer, M., Tscheschlok, K. and Volkmann, Ch. (2013) Full System Decontamination at German Nuclear Power Plant Unterweser. International Journal for Nuclear Power, 58, 216-220.

Overview for Improving Steam Turbine Power Generation Efficiency

Abolaji Joseph Omosanya[1], Esther Titilayo Akinlabi[1,2] and Joshua Olusegun Okeniyi[1,2]

[1]*Department of Mechanical Engineering Science, University of Johannesburg, Johannesburg, South Africa*
[2]*Department of Mechanical Engineering, Covenant University, Ota, Ogun State, Nigeria*

ABSTRACT

Electricity is an integral part of every society for which demand is growing continuously, whereas the production is still based on limited sources of energy derived mainly from steam and gas turbines, the turbomachinery. This paper presents an overview for preliminary study on the optimization of the design of the steam turbine. This was done with a special focus on the last stage low pressure turbine blades, for the reason that the design parameters of this component exhibit influence on the efficiency of power generation from the steam turbine electric power generating system. For supporting the study, a practical overview of the Egbin thermal power station, Nigeria, was included in the study with the parameters from the last stage low pressure turbine blade for this energy generation installation. By these, suggestions that could be undertaken for improving efficiency of the steam power plant for enhancing sustainability of electric power generation were also detailed in the paper.

Keywords: *Turbine blade; Efficiency; CFD*

1. INTRODUCTION

Man needs and uses energy at an increasing rate for his sustenance and well-being ever since he came on earth some million years ago [1-2]. The continued rise in World's population with the seemingly increase in demand for power has indisputably highlighted the need to increase power generation capacities. According to global statistics [3], energy supply has shown a steady increase in the past years In spite of these, however, it is still generally considered that the available energy generation plants are not sufficient to produce the required power [4-5], thereby necessitating needs to invest more in the construction of new power generation plants or seek alternative forms of energy generation. For remaining competitive, therefore, it is important for power generation companies and private power operators to seek ways for optimizing existing plants in order to improve efficiency and reliability, as well as to reduce the cost of plant operations and maintenance [6-7]. Although, there is renewed interest for sustainable, renewable and affordable energy technologies, the main energy producers still employ turbomachines, with the dominant primary source of energy being the fossil fuels, the versatile primary energy generation source globally [8-9]. However, environmental restrictions ensuing from emissions from fossil fuel combustions are driving researchers and stakeholders towards the search for more efficient electric energy generation systems [10-12]. Turbomachinery driven thermal power plants for generating electricity via fossil fuel can employ any of internal combustion plants, nuclear power plants, gas turbine plants, or steam power plants which amounts to about 80% of the electricity generated worldwide [11]. In this paper, the overview of steam power generating plant was deliberated upon with a view towards improvement of electric energy generation efficiency that could lead to cost saving and reduced environmental impacts.

2. STEAM POWER SYSTEM FOR ELECTRICITY GENERATION

2.1. The turbomarchinery and turbine energy generation system

The prominent forms of energy used in the industrial process for heating are electricity, directfire heating and steam. Heating using electricity basically involve the use of heating elements (normally resistors) which converts the electrical current flowing through it to heat energy. For direct-fire heating, hot gases from a burner transfer their heat energy to the system. Steam which is the predominantly used form of energy for process heating is also used for pressure control, mechanical drive, and is sometimes directly injected into the process as a source of water for process reactions. Steam is often used compared to other sources of energy for heating because of its performance advantages that include non-toxicity, ease of distribution, high heat capacity and low running and set-up cost.

The turbomachinery generally transfers energy between a fluid and a rotor. While a turbine transfers energy from a fluid to a rotor, a compressor transfers energy from a rotor to a fluid. The overall efficiency of the simple steam power plant is generally lower than that of other power plants such as the hydro, diesel and nuclear power plants. Therefore, the design engineer aims at producing a steam turbine characterized by an optimal energy conversion with an optimum efficiency. The overall efficiency of a steam turbine power plant solely depends on the performance and reliability of the turbine and its components. Thus any slight improvement can increase power availability, improve reliability, decrease equipment and component costs, and generate tangible operational savings. Taking into consideration the pronounced use of turbines in power generation, minor improvement in efficiency have both economic and environmental advantages for it will transform into billions of savings and result in reduced emission per unit of energy generation [6,13].

2.2. Efficiency improvements in steam turbine power plants

In the last three decades, engineers and designers have done several researches, simulations and experimentations on turbine systems to make industrial steam turbines more efficient and reliable [14]. These works have led to the improvement and the design of more powerful, efficient, and reliable steam turbines and the development of new materials and manufacturing processes. During the past years, the steam turbine inlet temperatures have been increased further for improving the efficiency of the steam plant, especially, with recent designs having up to 620°C inlet temperature even as this can still be increased in the near future [15]. This increment makes the review into the heat transfer characteristics of the materials chosen for the design of the turbine and most especially the turbine blades to be important and necessary. In addition, the structural integrity of all rotating components, mainly the rotor and the shaft is a key factor for successful, efficient and reliable operation of any turbomachinery. The integrity depends on the successful resistance of the moving parts of the turbine to the steady and alternating stresses imposed on them, both internally by the heat transferred and externally by the pressure from the steam. Due to the high cost of setting up a steam power generation plant, it is important to have an optimal design to maintain good returns on the capital investments by the power generation companies. Thus, the turbine has to be designed in such a way that overdesigning of the parts - such as blade and material selection - is avoided while fulfilling their basic requirement.

2.2.1. The energy cycle for the steam power turbine system

Figure 1 below shows the ideal Rankine cycle. Steam power plants are assumed to follow this ideal thermodynamic cycle for the explanation of their basic processes. It involves two isentropic and two isobaric processes. However, for practical operation, the working fluid is reheated and made to run through another turbine before going to the condenser. The thermodynamic processes through the states of the cycle shown in Figure 1 are:

- 1-2: Isentropic compression. Saturated liquid enters the pump at state 1 is pumped from low pressure to the pressure required by the boiler.
- 2-3: Isobaric heat addition. The saturated water in the boiler at the required pressure is heated up by an external heat source and leaves as superheated vapour.
- 3-4: Isentropic expansion. The superheated vapour expands in the turbine which is used to drive a generator to produce power.
- 4-1: Isobaric heat rejection. The vapour is condensed to saturated liquid in the condenser.

The turbine is the equipment responsible for the isentropic expansion which refers to the state 3 to 4 of the four reversible processes of the steam generation

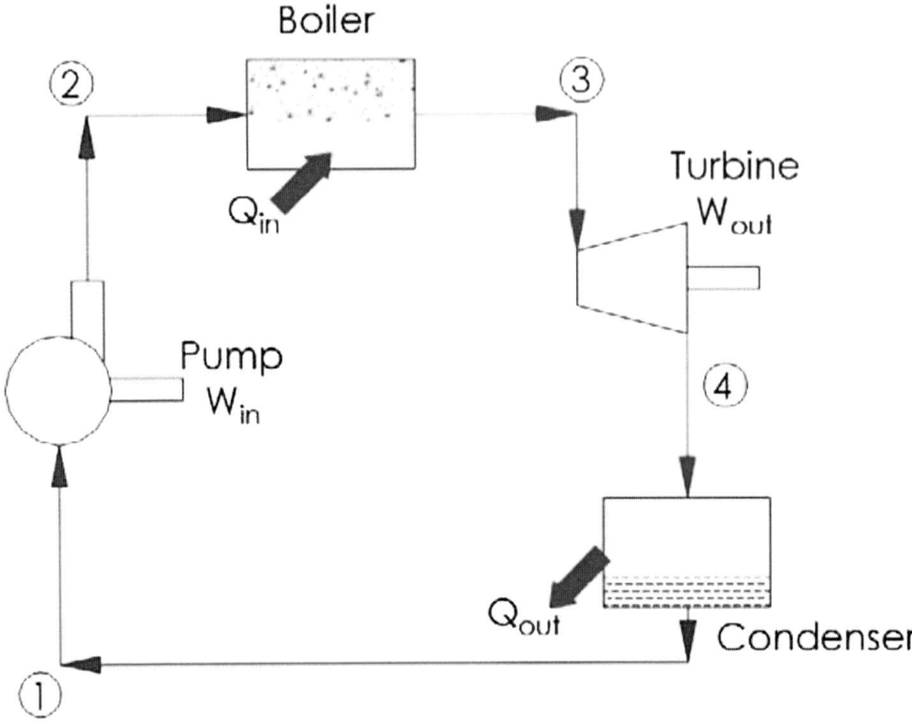

Figure 1. *The ideal Rankine cycle for a steam power generation plant*

plant, as shown from the ideal Rankine cycle of a steam power generation plant in Figure 1. For the steam generating plant design, the turbines are composed of several pairs of stationary (stator) and rotating (rotor) blade rows. The superheated steam with high velocity and pressure passes through the stationary blade rows also called nozzles and hits the rotating blades, sometimes referred to as buckets which are mounted on a shaft. The design of the nozzles and diaphragms of a turbine helps to drive the flow of the steam into a well formed, high velocity jets as the steam expands from inlet pressure through the various sections of the turbine blade to the exhaust pressure. This steam after passing through the nozzle produces a dynamic pressure on the blades installed on the rotor in which the blades and the shaft both start to rotate in the same direction [16]. The next arrangement of stationary blades helps to increase the velocity of the steam in a circumferential direction towards the next set of rotary blades. In a steam turbine, the mechanical energy in form of pressure present in steam is extracted and converted into kinetic energy by allowing the steam to flow through the stationary blades. This kinetic energy is transmitted to the moving blades connected to the alternator of a steam turbine generator. Turbine generator converts the mechanical energy from the rotor in form of rotary motion into electrical energy.

The turbine blade is the most important element in the steam power generation system as it transmits the pressure energy from the steam to the rotor.

Turbine blades can be classified based on their interaction with the steam as impulse or reaction blades. It is important to note that turbine blades are not exactly the same throughout the section of the steam turbine. Steam turbines are designed with multi steam expansion stages, which are basically high, intermediate and low pressure stages [9]. In this paper, emphasis will be made on the low pressure (LP) turbine blade, which is in the last stage of the turbine and that is, characteristically, the longest blade system of the turbine.

2.2.2. Factors affecting efficient steam power turbine blade design

The LP, which is also the last stage, turbine blade is characterized by a number of issues due to the length of its blades. These issues, among others, include large centrifugal stress, low rigidity and high Mach number flow. The overall thermal efficiency of the steam turbine as well as its size and total power output significantly depend on the last stage blades, the reason for which designing an optimal range of last stage blades is important [17].The life span of the turbine blades is influenced by a number of parameters. Unlike blades of a compressor, turbine blades experiences harsh forces acting within its environment ranging from high operating temperatures to high varying blade load. Other factors affecting the life span of the turbine blade include flow mismatch during water and steam ejection at the extraction ports, partial or imperfect vacuum, water injection, incorrect steam conditions, presence of impurities in the steam, etc. During the design of the turbine blade, it is required of the design engineer to [18]:

- Make a choice of the mechanical and steam loading of the steam turbine taking each individual blade into consideration;
- Select the shroud and root configuration for each stages of the turbine;
- Determine the stiffness required for the rotor root;
- Make material choice based on hardness, machinability, surface finishing; as well as
- Determine the tightness of the installation.

It is these listed considerations that necessitate the importance of having to strike a balance between the conflicting requirements of low cost, good aerodynamic performance, and long life, when designing a blade for the steam power generating turbine system. Therefore, the optimization of these design choices will therefore go a long way towards improving steam power generation efficiency by the turbine system in an electric power generation plant.

3. A PRACTICAL OVERVIEW: THE EGBIN POWER PLANT, NIGERIA

For a practical overview, this paper examines the steam power turbine system for electric enery generation at the Egbin thermal power plant, Nigeria. The plant is located at Ikorodu, Lagos State, Nigeria [13]. The plant, which has six turbine stations, was commissioned in 1987. Each individual station operates on a reheat-regenerative cycle and can be fired using either gas or heavy oil. The construction of the plant commenced in 1984 under the supervision of the Japanese and French

Table 1. *Operational parameters of the last stage low pressure turbine blade from Egbin power plant*

Parameter	Unit	Value
Rotational speed	rpm	3000
Blade length	mm	750
Chord length	mm	45
Reynolds number	dimensionless	4.9×10^5
Velocity	m/s	34.14
Coefficient of viscosity	kg/ms	1.009×10^{-5}

contractors with most of the equipment, including the Hitachi turbine being used at the plant, were sourced from Japan. The budget estimate of the plant was about 1 billion USD at that time while the life span of the plant was estimated at 25 years. According to [19], efficiency of the plant was averaged at 34.67% under the review period of 2000 to 2010 . However, in 2016 , the plant recorded an overall efficiency of 29% based on the output/input method of calculation [20]. This shows a significant drop in the efficiency of the power station, and thus, necessitates the needs for reviewing the system for improvement possibilities. In this study, the analysis for improvement being proposed is being carried out on the turbine blade. To improve the efficiency of the blade rows, the flow pattern of individual stage will need to be redesigned considering all the fluid forces acting on the end-wall contour of the steam turbine and the blade profile [21].

The Egbin power station, like any other steam power plant, works on the principles of the Rankine reheat regenerative cycle. The most efficient cycle for thermodynamic operation is the Carnot cycle but it is not practicable for steam power cycle because it is difficult to design a compressor which could handle working fluid that is in both liquid and vapour state. It is also difficult to control the condensation process to produce the working fluid in the desired state for compression.

Figure 2 shows the image of the last stage low pressure turbine blade currently in use at Egbin thermal power plant. The operational parameters of this turbine blade are as presented in Table 1.

Considering that the operational parameters of the last stage low pressure turbine blade system, shown in Figure 2, will exhibit dependencies on the blade length, aerofoil design and steam exhaust characteristics, it could therefore be inferred the efficiency of the steam turbine electric power generation plant could be improved by:

- Increasing the length of the low pressure last stage turbine blades that will lead to reduced exhaust loss, and which, in turn, will bring about increase in the capacity of the turbine, thus, improvement to the efficiency of the steam turbine plant. This, when viewed broadly reduces the cost of construction against power/capacity of the turbine.
- Optimizing the aerofoil design and improving the angle of attack. The profile of the aerofoil determines how effectively the kinetic energy of the steam will

Figure 2. *Last stage low pressure turbine blade*

be converted to mechanical energy. An optimized design of the aerofoil will ensure a better percentage of the kinetic energy of the steam is used in the turbine.

- Enlarging the steam exhaust at the low pressure turbine. Steam expands more in the low pressure turbine, hence, the long turbine blades. If the exhaust of the low pressure turbine is not wide enough to accommodate the rate of expansion, back pressure will be experienced by the blades which will inhibit the overall efficiency of the power plant.
- Conducting further research into the materials properties of the low pressure turbine blade for steam turbine power generating system, especially, towards the improvement of efficiency that could be induced from improved turbine materials properties.

4. CONCLUSION

An overview for improving steam power turbine energy generation efficiency has been carried out in this work. The conclusion following from the study includes:

- Sustainability for the purpose of lower cost, leading to economical generation of electric power, and environmental effects via reduced emission per unit of energy generation, by using the steam power generation plant will require improving the system efficiency.
- It could be possible to improve efficiency of the steam turbine power plant system through optimization of the design parameters of the last stage LP turbine blade.
- It will also be necessary to conduct further research into the materials properties of the LP turbine blade for steam power generation system for the improvement of efficiency improvement that could be accrued from improve turbine materials properties.

Acknowledgements

The authors wish to acknowledge the support offered by Egbin thermal power station in actualization of this research work for publication.

REFERENCES

1. Okeniyi, J. O., Moses, I. F. and Okeniyi, E. T. (2015). Wind characteristics and energy potential assessment in Akure, South West Nigeria: Econometrics and policy implications. International Journal of Ambient Energy, 36(6), 282-300.
2. Ajayi, O. O., Fagbenle, R. O., Katende, J. and Okeniyi, J. O. (2011). Availability of wind energy resource potential for power generation at Jos, Nigeria. Frontiers in Energy, 5(4), 376 – 385.
3. The International Energy Agency (IEA) (2018). Total Primary Energy Supply (TPES) by source: World 1990-2016. IEA Key energy statistics https://www.iea.org/statistics/?country=WORLD&year=2016&category=Energy%20supply&indicator=TPESbySource&mode=chart&dataTable=BALANCES Accessed: July 11,2019.
4. Okeniyi, J. O., Atayero, A. A., Popoola, S. I., Okeniyi, E. T. and Alalade, G. M. (2018). Smart campus: data on energy generation costs from distributed generation systems of electrical energy in a Nigerian University. Data in Brief, 17, 1082-1090.
5. Okeniyi, J. O., Anwan, E. U. and Okeniyi, E. T. (2012). Waste characterisation and recoverable energy potential using waste generated in a model community in Nigeria. Journal of Environmental Science and Technology, 5(4), 232-240.
6. Ray, T. K., Datta, A., Gupta, A. and Ganguly, R., 2010. Exergy-based performance analysis for proper O&M decisions in a steam power plant. Energy Conversion and Management, 51(6), 1333-1344.
7. Mahamud, R., Khan, M. M. K., Rasul, M. G. and Leinster, M. G. (2013). Exergy analysis and efficiency improvement of a coal fired thermal power plant in Queensland. In M. Rasul, Thermal Power Plants - Advanced Applications. IntechOpen. DOI:10.5772/55574

8. Ajayi, O. O., Fagbenle, R. O., Katende, J., Aasa, S. A. and Okeniyi, J. O. (2013). Wind profile characteristics and turbine performance analysis in Kano, north-western Nigeria. International Journal of Energy and Environmental Engineering, 4(1), 27.

9. Rosen, M. A. and Tang, R. (2008). Improving steam power plant efficiency through exergy analysis: effects of altering excess combustion air and stack-gas temperature. International Journal of Exergy, 5(1), 31-51.

10. Okeniyi, J.O., Loto, C.A. and Popoola, A.P.I. (2014). Morinda lucida effects on steelreinforced concrete in 3.5%NaCl : Implications for corrosion-protection of wind-energy structures in saline/marine environments. Energy Procedia, 50, 421-428.

11. Jonshagen, K. (2011). Modern Thermal Power Plants-Aspects on Modelling and Evaluation. Doctoral dissertation, Department of Energy Sciences, Lund University, Sweden.

12. Ajayi, O. O., Fagbenle, R. O., Katende, J., Okeniyi, J. O. and Omotosho, O. A. (2010). Wind energy potential for power generation of a local site in Gusau, Nigeria. International Journal of Energy for a Clean Environment, 11(1-4), 99-116.

13. Eke, M. N., Onyejekwe, D. C., Iloeje, O. C., Ezekwe, C. I. and Akpan, P. U. (2018). Energy and exergy evaluation of a 220MW thermal power plant. Nigerian Journal of Technology, 37(1), 115-123.

14. Singh, M. P. and Lucas, G. M. (2011). Blade Design and Analysis for Steam Turbines. The McGraw-Hill Companies, New York.

15. Fadl, M., Stein P. and He, L. (2017). Full conjugate heat transfer modelling for steam turbines in transient operations. International Journal of Thermal Sciences, 124, 240-250.

16. Bloch H. P. and Singh M. P. (2009) Steam Turbines; Design, Applications, and Rerating. The McGraw-Hill Companies, New York.

17. Senoo, S., Ono, H., Shibata, T., Nakano, S., Yamashita, Y., Asai, K., Sakakibara, K., Yoda, H. and Kudo, T. (2014). Development of titanium 3600rpm-50inch and 3000rpm 60inch last stage blades for steam turbines. International Journal of Gas Turbine, Propulsion and Power Systems, 6(2), 9-16.

18. Naumann, H. G. (1982). Steam turbine blade design options: how to specify or upgrade. In P. E. Jenkins, Proceedings of the 11th Turbomachinery Symposium. Turbomachinery Laboratories, Texas A&M University.

19. Adegboyega, G. and Odeyemi, K. (2011). Performance Analysis of Thermal Power Station; Case Study of Egbin Power Station, Nigeria. International Journal of Electronic and Electrical Engineering, 281-289.

20. Egbin-Power, (2016). Securing Our Future; 2016 Sustainability Report. Egbin-Power, Lagos. Www. http://egbin-power.com/wp-content/uploads/2018/02/2016-EgbinSustainability-Report-Spread.pdf, Accessed July 11, 2019.

21. Saito, E. I. J. I., Matsuno, N. A. R. I. Y. U. K. I., Tanaka, K. E. I. Z. O., Nishimoto, S. H. I. N., Yamamoto, R. Y. U. I .C. H. I. and Imano, S. H. I. N. Y. A. (2015). Latest technologies and future prospects for a new steam turbine. Mitsubishi Heavy Industries Technical Review, 52(2), 39-46.

CHAPTER 6

Application of Probabilistic Model for Marine Steam System Failure Analysis under Uncertainty

Sidum Adumene and Samson Nitonye

Department of Marine Engineering, Rivers State University, Port Harcourt, Nigeria

ABSTRACT

In ship and offshore operations, machinery systems have associated operational hazard because of the prevailing harsh environment. Therefore, the need for an overall evaluation of the associated risk and failures of these systems, such as the marine steam boiler, is crucial to the industry. The concept of probability risk model is used to model the failure mode considering the overall risk associated with the system as a whole. The rate of occurrence of the failure that described the basic events as represented by the fault tree was developed to model the marine steam system. This specific event was implemented and evaluated to estimate the failure frequencies of the overall systems, based on the available failure rate in core literatures. A risk model which is hazard severity weight with its failure frequencies, and the time of operation was applied in the analysis. The probability of failure of the boiler system was estimated at 0.323225 at 35,040 operating hours with hazard severity weight of catastrophic if it occurs. The associated failure frequency calculated for the period is 1.114×10^{-5}*. The over failure frequency of the marine steam system for the period of consideration is conditioned on the pre-defined minimum cut sets of the top event. This therefore agreed with the fact that the basic events with their failure frequencies will lead to the catastrophic failure of the entire system within the period if the maintenance plan is not proactive.*

Keywords: *Machinery Operation; Marine Steam System; Boiler; Risk; Probability; Failure Mode*

1. INTRODUCTION

Marine steam systems are units that generate steam for electric power generation or process heating and operation onboard and other offshore operations. On board ship and in process plant, steam is used for personal uses such as producing heat and hot water. Marine steam system such as boilers contains two basic systems [1], such as the steam water system and the fuel air-flow gas system. Fuel and air served as inputs system, and the process of combustion is done in the wind-box.

The boiler outputs systems are the flow gas and ash systems [2]. On board merchant ships, several regulations are required to effectively secure every dangerous part of a ship's machinery to ensure safety of personnel and the ship. Safety remains a human issue, and a human solution has to be found in solving or preventing safety problems [3].

A ship machinery space or engine room is the compartment of the ship where the main engine(s), generator(s), compressor(s), pump(s), boiler plants and other major machineries are located [4]. The ship machinery space where the boiler is located is the most hazardous area of the ship. Despite the fact that marine operations have tended to rely on meeting regulatory requirements, industry codes of practice, or Classification Society Rules, taking all precaution and safety measure while handling engine room machinery system, accident are bound to take place in the ships engine room. There exist specific requirements for machinery in the engine room regarding starting and stopping of the plant. Therefore, strict precaution is needed for boiler system being one of the most important systems on board ships, due to its multipurpose functionality. Although machinery failure is dangerous, some are more dangerous than the other. Therefore, prompt actions on the matter of risk and failure control are crucial in the operation of marine steam systems [5].

Risk-based analysis is used as a monitoring object (condition monitoring, diagnostics and servicing) that integrates technical, economic and safety issues together to provide solution to system problem [6]. Ship operational risks have not been properly defined because of the system complexity, especially in multipurpose carriers, containership. Its evaluation depends on design and construction assumptions, which are burdened with some indetermination of construction solutions. The interaction of hydro meteorological conditions in the ship operating environment, plant maintenance, crew expertise and their habits are contributory factors to the failure or degradation of power system onboard [7]. Risk-based analysis of onboard energy system considers data which includes design data, operational data/efficiency, failure analysis, cost implementation for servicing and repairs.

In reference to offshore vehicle as a safe maritime facility, different operation strategies and maintenance methods can be applied to the sub-system to sustain reliability and performance [7]. Therefore, for safe performance and sustainable reliability, it is necessary to ensure flow of mechanical and electrical energy, as well as heat for the sea voyage, loading function, steering and commination utilization. This greatly depend on the main power system consisting the safety of a ship power system in its real operating conditions, the technical, design and operating parameters are used. The general operating structure as well as sea condition directly influence the required state of any given power system onboard ship[8].

This research seeks to investigate associated risk and failure mode of a steam system that form major part of a marine propulsion system and also for process offshore operations. There exists little research in the area of modeling boiler system risk in the maritime industry and none have applied a probabilistic based technique to predict the trend of the system performance within its design and off-design envelopes. Hence, we proposed the application of a predictive probabilistic tool to model the entire boiler system with the aim to ascertain its failure rate, frequency and consequences for a predefined duration. This provide a novelty application of the probabilistic model in predicting failure characteristic of a marine steam system (boiler plant) which previous literatures did not provide a holistic illustration as is done in this research.

2. DESCRIPTION OF BOILER SYSTEM OPERATION FOR SEA GOING VESSELS

The boiler system is configured with feed water drum and the water utilizes the heat energy released by the burning fuel. This energy gained converted the water into the form of steam with very high temperature and pressure. The system has a combustion chamber for the pressurization of the fuel to high temperature and air is supplied to this combustion chamber through a separate arrangement to enhance combustion [9].

Heat exchange occurred from the hot gas to the water through the boiler drum wall of a given large surface area, which enables the highest rate of energy transfer [10]. The energy from the burning fuel is then used for different purpose onboard, such as:

- Steam production for process,
- Steam superheating for power generation.

2.1. Types of Boiler

We have two main types of boiler. These two are the basic, all other boiler are different versions of them.

- Water tube boiler,
- Fire tube boiler.

The water tube boiler is a shell and tube heat exchanging system where the exhaust gases are the product of combustion passes over the tubes containing flowing water. Research shown that boiler tubes are made of materials that typically withstand higher internal pressure compare to large chamber shell in a fire tube boiler. For higher temperature application, water tube boilers are mostly used with high steam pressures (as high as 3000psi) are required. Water tube boilers performance also show high efficiencies and can generate saturated or superheated steam as the need arises. The merit of water tube boiler to generate high pressure steam (superheated steam) makes it attractive in applications for steam turbine power generation [10].

In Fire tube boiler, heat is transferred from the combustion gases pass inside boiler tubes, to the water on the shell side. Fire tube boilers are described by their number of passes configuration. This describes the number of times the combustion (or flue) gases flow the length of the system as they transfer heat to the water [11]. The passes are arranged in a counter-flow pattern. For design configuration, the gases turn 180 degrees and pass back through the shell.

2.2. General Boiler Hazard

Over the year risks associated with boiler plant operations has been drawn from the following hazardous situations within the boiler as outline below.

- Control system malfunction;
- Fire;
- Fire side explosion;
- Loss of power supply;
- Loss of water;
- Overpressure;
- Overheating (overheating as a result of low water is the most common cause of boiler damage or explosions, usually a result of the malfunction of the automatic controls);
- Unauthorized access;
- Unauthorized modifications and repairs.

The overall operation and maintenance of boiler plant gives rise to a high level of risk, basically the super heater, boiler water level, boiler furnace fuel and air supply, boiler safety valves and more will be analyzed in this research work. Safety and risk are related. The safer operation is defined as a case of fewer risks. Property damage is considered a risk that might cause injury or loss of life [12]. Current regulations demand the implementation of safety requirements in boiler operation to keep risks under safe control. Therefore, the importance of risk and safety analysis in the operation of ship machinery is key to the safe manning of ships. We have occurrences that can result to failure of steam steams, these includes:

- Melt Down

Melting down occur when the heating surface metal reach its melting point, which is temperature dependent. This occurs mostly when boiler operating at a very low water level. Although it not results to boiler explosion in itself, but it effects causes major damage to the boiler and create a dangerous situation which could lead to an explosion.

- Thermal Shock

This is a condition where low water causes the heating surfaces to become overheated and then cooler water is added. The water then flashes to steam which expands 1600 times its volume as water and causes the explosion because there is not enough room for the steam to expand.

- Combustion Explosions

These can be a result of gases which build up and an ignition source ignites the gases. This can happen inside the boiler or outside. There are safety devices in place to avoid these situations and we will discuss these in the following slides.

- Steam Pressure

Excessive steam builds up which exceeds the design pressures of the vessel. There is also safety device to prevent this.

3. OPERATIONAL FRAMEWORK FOR RISK AND SAFETY-BASED ASSESSMENT

Power system functionality and effectiveness form a strong hold on the seaworthiness of all oceans going vehicle. Ship power system management involves planning and decision making, organizing, managing and control. This system management from the operational strategy which defines the methods of maintaining the technical condition of an assumed level as in the required time of their operation, often estimated with key indication of effectiveness [7]. This framework entailed Control Based Maintenance (CBM), Planned Maintenance Strategy (PMS), Preventive Maintenance (PM) Reliability Centered Maintenance (RCM) and Time-Based Maintenance (TBM).

Operating maintenance management of ship power system should be carried out in compliance with regulation of the ISM Code, classification solution and the applied operating strategy guaranteeing that the power system will perform the task facing it. The operational approach is justified by the creations of maintenance strategies of technical conditions of ship equipment based on optimizing economic resources on technology and safety [7]. Although control system with failure analysis are been implemented to a limited extent to explain risk-based analysis. In Nwaoha et al. [13], they combined fuzzy logic, ER, and AHP algorithms to assess the propeller operational safety and reliability. The result should degree of confidence in the application of these tools. These provide the categorization of the various threats to the safe operation of the propeller as well as the cost implication on the ship owner. Many maintenance strategies lack reliability base on technical risk. Most times, boiler system failure is one such dangerous accident which is caused because of the following reasons:

- Fuel dripping inside the furnace of the boiler (blowback and even explosion),
- Misfiring,
- Overheating of boiler due to loss of water circulation,
- No pre- and post-purging,
- Exhaust gas boiler fire.

3.1. Fault Tree Application for Failure Analysis

Many safety assessment approaches, such as probabilistic risk assessment method, have been widely used but do have some challenges such as reliance on the failure rate data, which may not be available. The fault tree analysis (FTA) is a productive hazard analysis technique widely used in the maritime industry [14]. According to Nwaoha et al., [14], FTA is carried out using deductive analysis from the top event, which is the undesired event followed by causal relationships of the failures leading to that event identified by experience from previous accident of the report in question [14]. FTA can be evaluated using both qualitative [15] [16] and quantitative techniques [15] [17]. Similarly, the reduction of FTA can be carried out by the Boolean algebra method [16] [18] and the BDD method [19] [20]. It is important to understand that a fault tree is not a model dealing with all possible system failures and it covers the most credible faults as assessed by the analyst [18]. FTA uses different types of gates for it construction, which makes it a static or dynamic fault tree [21], a non-coherent [22] or coherent fault tree. A coherent fault tree which is mostly used in risk assessment in the marine industry, uses OR and AND gates to construct its tree [14]. Risk modeling of boiler systems in this case, we used FTA because of it compatibility for effective cost modeling and hazard consequences.

Wang and Trbojevic [18] defined Risk assessment as a comprehensive estimation of the probability and the degree of the possible consequences in a hazardous situation in order to select appropriate safety measures. Risk assessment is a useful tool for both the marine and process industry and it can be used to identify areas that need regular maintenance and repair for performance sustainability. Mathematically, risk can be expressed as follows [14]:

$$\text{Risk} = \text{consequences} \times \text{likelihood} \tag{1}$$

$$= \text{Hazard severity } (s) \times \text{ failure probability } (p) \tag{2}$$

$$= \text{Hazars severity's weight } (S_w) \times \text{ failure probability } (p) \tag{3}$$

In the risk assessment of the marine boiler, failure probability intends to define and follow an exponential distribution path, such that;

$$P = 1 - e^{-\lambda t} \tag{4}$$

$$\lambda = \frac{-\ln(1 - P)}{t} \tag{5}$$

Therefore

$$R = S_w \times \left(1 - e^{-\lambda t}\right) \tag{6}$$

where

$1 - e^{-\lambda t}$ is the exponential distribution formulas,

P is the failure probability,

λ is the failure rate or frequencies,

t is the time of interest.

The basic risk level (basic events) of the whole boiler system can be determined by the sum of the risk associated with its systems.

$$R_T = R_{\text{subsystem}(1)} + R_{\text{subsystem}(2)} + \cdots + R_{\text{subsystem }(n)} \tag{7}$$

where R_T is the total risk of boiler system,

$R_{\text{System }(i)}$ = risk of the boiler subsystem $i, i = 1, 2, \cdots, n$ or $(i \in n)$.

Therefore, by substitution we have the following

Table 1. *Boiler system hazard consequence.*

Severity Index	Description	Equipment	Personnel
4	Catastrophic	System loss	Death
3	Critical	Major system damage	Severe injury/illness
2	Marginal	Minor system damage	Minor injury/illness
1	Negligible	Non-significant damage	Non-significant injury/illness

Table 2. *Hazard probability [14].*

Level	Description	Equipment
A	Frequent	Likely to happen
B	Probable	Several times during lifetime
C	Occasional	Likely to happen once
D	Remote	Unlikely but possible during lifetime

$$R_T = S_{w1} \times \left(1 - e^{-\lambda_1 t_1}\right) + S_{w2} \times \left(1 - e^{-\lambda_2 t_2}\right) + \cdots + S_{wn} \times \left(1 - e^{-\lambda_n t_n}\right) \quad (8)$$

where

$S_{w(i)}$ is the hazard severity's weight of the boiler subsystems i,
λ_i is the failure rate of the boiler subsystem i,
t_i is the time interest of the boiler subsystem i,
$i = 1, 2, 3, \cdots, n$ or $(i \in n)$.

The associated risk which defined the top event of an FTA modeling of boiler system is evaluated using the level/consequences pathway.

Qualitative risk analysis is used identified hazard that can be categorized to be catastrophic, critical, marginal and negligible categories as shown in Table 1. The occurrence probability of the hazards can be expressed as frequent, probable, occasional or remote as the case may be as shown in Table 2. This analysis tends to obtain the level/consequence of risk in the early stage where data are not available for quantitative risk analysis. It is used to analyse each individual component by applying characteristic features such as failure rate, repair rate, system logic, maintenance schedule, mission time and human error. These characterized features are used to formulate a mathematical model which helps to identify high risk areas needed to be controlled. The risk associated with the boiler system and their combined consequence can be represented using combination therapy as indicated in Table 3. Different actions are to be taken to eliminate or control the hazards associated with the plant. These are group as follows [14]:

Table 3. *Boiler system risk assessment matrix [14].*

Hazard Severity	Weight	A (Frequent)	B (Probable)	C (Occasional)	D (Remote)
Catastrophic	1000	A-01	B-01	C-01	D-01
Critical	100	A-02	B-02	C-02	D-02
Marginal	10	A-03	B-03	C-03	D-03
Negligible	1		No significant hazards		

- Adequate design and operational actions are required to eliminate or control hazards classification as A-01; A-02; A-03, B-01, B-02 and C-01;
- Hazard consequences must be controlled for classification B-03; C-03, and D-01;
- Hazard control is desirable if cost effective for hazard classified as C-03 and D-02;
- Hazard control is not cost effective for hazards classified as D-03.

3.2. Hazard Identification in Marine Boiler Operations and Probabilistic Analysis

Potential hazards associated with the boiler system can be identified by experts in the maritime industry. Wang and Trbojevic [18] defined hazard as a physical situation with a potential to cause injuries/deaths, property damage to the environment, or some combination of these. HAZID is carried out from systematic reviews of all operational modes modeling different sections of the boiler plant [14]. The team ensures that the process is proactive and not confirmed only to hazards that have materialized in the past operation. The following hazards were identified on investigation.

A system failure may occur unexpectedly and care is needed for supercritical equipment like the marine steam system. The boiler system failure may be rooted in structural defects, corrosion, stress rupture, fatigue, erosion, and lack of quality control [23]. Stress rupture occurs when the stress cracks and creep set in simultaneously, as a result of thermal overheating that may be long or short terms. This is predisposed by excessive pressure and temperature to pressure relief valve and temperature sensor failure. Fatigue occurs when vibration, thermal stress, and corrosion effect simultaneously set in. These predisposing factors include structural defects, pressure relief valve failure and corrosion [24]. Corrosion causes are critical damage and deformation on boiler material structure, and this occurs as a result of fuel ash, embrittlement, carbide graphitization, dew point, pitting, oxidation and intergranular. Erosion occurs by influence of fly ash, falling slag, soot blower and fuel ash. Lack of quality controls can result from the design, fabrication, operation and maintenance, oxidation, chemical excursion, weld defect and structural defect. Structural defects, temperature sensor failure, pressure relief valve failure, personnel faults, impurities, corrosion are basic events as shown in Table 4 and Figure

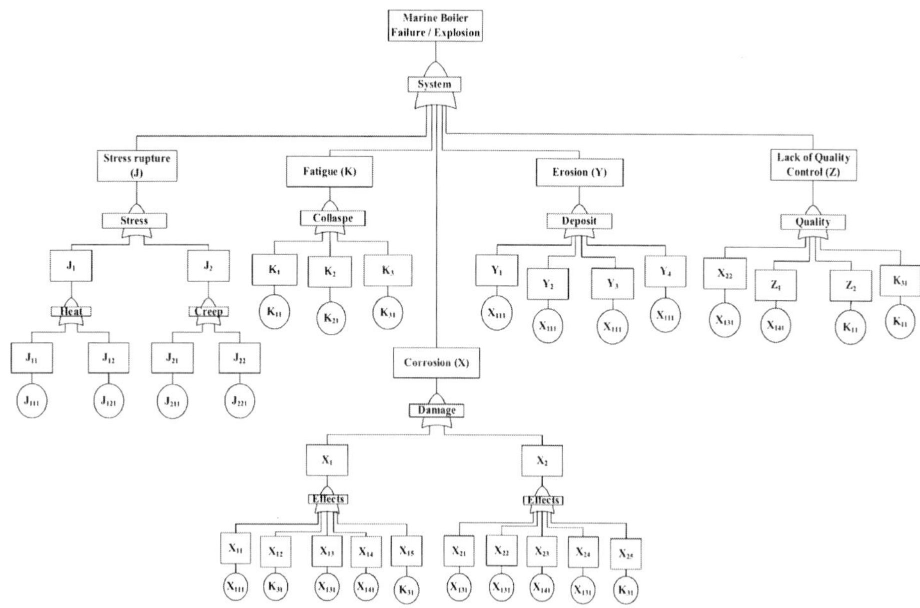

Figure 1. *Fault tree of marine steam system failure mode.*

1. The Figure 1 is drawn from the categorization of the prevailing failure mode associated with the case study.

For the quantification of the top event, the marine boiler has five major events with different basic events and their failure consequences. Although the overall basic events are grouped into eight for the purpose of this research and their assigned frequency of failure and probability is shown in Table 5.

For the entire marine boiler system, the failure probability gives

$$
\begin{aligned}
&P(\text{ Marine boiler system }) \\
&= P(A) + P(B) + P(C) + P(D) + P(E) + P(F) + P(G) + P(H) \\
&\to P(A) + P(B) + P(C) + P(D) + P(E) + P(F) + P(G) + P(H) \\
&- P(A) \cdot P(B) - P(A) \cdot P(B) - P(A) \cdot P(C) - P(A) \cdot P(D) \\
&- P(A) \cdot P(E) - P(A) \cdot P(F) - P(A) \cdot P(G) - P(A) \cdot P(H) \\
&- P(B) \cdot P(C) - P(B) \cdot P(D) - P(B) \cdot P(E) - P(B) \cdot P(G) \\
&- P(B) \cdot P(H) - P(C) \cdot P(D) - P(C) \cdot P(E) - P(C) \cdot P(F) \\
&- P(C) \cdot P(G) - P(C) \cdot P(H) - P(D) \cdot P(E) - P(D) \cdot P(F) \\
&- P(D) \cdot P(G) - P(D) \cdot P(H) - P(E) \cdot P(F) - P(E) \cdot P(G) \\
&- P(E) \cdot P(H) - P(F) \cdot P(G) - P(F) \cdot P(H) - P(G) \cdot P(H)
\end{aligned}
$$

Table 4. *Risk/hazard events algorithm coding for marine boiler system.*

Major Events	Basic Events	Coding
	Stress Cracks	J1
	Thermal Overheating	J11
	Temperature Fluctuation	J12
	Sensor Failure	J111
	Relief Valve Failure	J121
Stress Rupture (J)	Creep Failure	J2
	Overheating	J21
	Long-Term Overheating	J211
	Short-Term Overheating	J212
	Pressure Relief Valve Failure	J2111
	Temperature Sensor Failure	J2121
	Vibration	K1
	Defects	K11
	Thermal Stress	K2
Fatigue (K)	Temperature Sensor Failure	J2121
	Corrosion Effect	K3
	Material/Structural Defect	K31
	Fireside Corrosion	X1
	Fuel Ash	X11
	Personnel Faults	X111
	Embrittlement	X12
	Sulfidation	X13
	General Impurities	X131
Corrosion (X)	Nucleate Boiling	X14
	Material Impurities	X141
	Carbide Graphitization	X15
	Waterside Corrosion	X2
	Dew Point	X21
	Oxidation	X22
	Formation of Sigma Phase	X23
	Pitting	X24
	Intergranular	X25
	Fly Ash	Y1
Erosion (Y)	Falling Slag	Y2
	Soot Blowers	Y3
	Fuel Ash	Y4
	Oxidation	X22
Lack of Quality Control (Z)	Chemical Excursion	Z1
	Weld Defects	Z2
	Material/Structural Defects	K31

Table 5. *Marine boiler system basic events and their failure frequency and probability.*

Basic Event	Failure Frequency	Failure Probability [14]
Structural defects	$2.31e^{-006}$/h [25]	P[A]
Corrosion effects	$1.115e^{-006}$/h [25]	P[B]
Pressure relief system	$2.12e^{-005}$/h [25]	P[C]
Fire and explosion	$1.78e^{-006}$/h [25]	P[D]
Overpressure	0.01/h [26]	P[E]
Material defect	$11.15e^{-06}$/h [25]	P[F]
Sensor failure	0.03/h [27]	P[G]
Overheating (rupture)	$2.96e^{-010}$/h [26]	P[H]

And if the boiler is estimated for a 20 years period and is subjected to major failure analysis every four (4) years period, we have

At $t = 35,040$ h,

$$P(A) = 1 - \mathrm{e}^{-\lambda t} \text{ where } \lambda = 2.31\mathrm{e}^{-006}$$

$$P(A) = 1 - \mathrm{e}^{-2.31\mathrm{e}^{-006 \times 35040}} = 1 - 0.922 = 0.078$$

$$P(B) = 1 - \mathrm{e}^{-1.115\mathrm{e}^{-006 \times 35040}} = 1 - 0.962 = 0.038$$

$$P(C) = 1 - \mathrm{e}^{-2.12\mathrm{e}^{-005 \times 35040}} = 1 - 0.928 = 0.072$$

$$P(D) = 1 - \mathrm{e}^{-1.78\mathrm{e}^{-006 \times 35040}} = 1 - 0.940 = 0.061$$

$$P(E) = 1 - \mathrm{e}^{-0.01 \times 35040} = 1 - 0 = 1$$

$$P(F) = 1 - \mathrm{e}^{-11.15\mathrm{e}^{-06 \times 35040}} = 1 - 0.677 = 0.323$$

$$P(G) = 1 - \mathrm{e}^{-0.03 \times 35040} = 1 - 0 = 1$$

$$P(H) = 1 - \mathrm{e}^{-2.92\mathrm{e}^{-010 \times 35040}} = 1 - 0.999 = 0.001$$

$P($ Marine boiler system$)$
$= 0.078 + 0.038 + 0.072 + 0.061 + 1 + 0.323 + 1 + 0.001 - (0.078 \times 0.038)$
$-(0.078 \times 0.072) - (0.078 \times 0.072) - (0.078 \times 0.061) - (0.078 \times 1)$
$-(0.078 \times 0.323) - (0.078 \times 1) - (0.078 \times 0.001) - (0.038 \times 0.072)$
$-(0.038 \times 0.061) - (0.038 \times 1) - (0.038 \times 0.323) - (0.038 \times 1)$
$-(0.038 \times 0.323) - (0.038 \times 1) - (0.038 \times 0.001) - (0.072 \times 0.061)$
$-(0.072 \times 1) - (0.072 \times 0.323) - (0.072 \times 1) - (0.072 \times 0.01)$
$-(0.061 \times 1) - (0.061 \times 0.323) - (0.061 \times 1) - (0.061 \times 0.001) - (1 \times 0.323)$
$-(1 \times 1) - (1 \times 0.001) - (0.323 \times 1) - (0.323 \times 0.001) - (1 \times 0.001)$
$= 0.323225$

$P(\text{ Marine boiler system }) = 1 - e^{-\lambda_1 \times 35040}$
$e^{-\lambda_1 \times 35040} = 1 - 0.323225$
$-\lambda_1 \times 35040 \times \ln e = \ln 0.6768 = -0.39038$
$\lambda_1 = 1.114 \times 10^{-5}$

The probability of failure of the boiler system is 0.323225 at 35,040 operating hours with hazard severity weight of catastrophic if it occurs. The associated failure frequency λ_1 calculated for the period is 1.114×10^{-5}.

The over failure frequency of the marine steam system for the period of consideration is conditioned on the minimum cut sets of the top event. This therefore agreed with the fact that when the basic events occur with their failure frequencies, it will lead to the catastrophic failure of the entire system if the maintenance plan is not proactive.

4. CONCLUSIONS

The process of assessing the associated failure trend in a marine steam system is crucial for every offshore operation. This assessment provides knowledge for improving the level of safety (reduction of risk) in boiler operation. The research focuses on the areas of high risk in steam supply and the major causative events such as stress rupture, fatigue, corrosion, erosion and poor-quality control.

The total risk of the system was analysed mathematically from the probabilistic model. The model estimates the various safety levels of the failure mode by setting up severity and consequences based on the prevailing events. The captured safety level of the subsystems was integrated into the model and the risks were ranked using the prevailing operating condition of the system. The sub-events for the purpose of this research were integrated under the five major events with their failure frequencies from existing literatures. The result shows that the adopted probabilistic model can be used for modeling the failure and risk associated with marine steam systems. This research did not consider the cost implication for the maintenance of the failed subsystems of the overall steam system. This research is not exhaustive. Further work can be done by employing fuzzy based model and evidential reasoning to analyse the professional perfectives on the relative severity classification and modeling.

Acknowledgements

We sincerely acknowledged Mr. Ezenwoke Anwurike Roy of the Department of Marine Engineering on his contribution to the improvement of this research work.

REFERENCES

1. Edward, A.C., Adumene, S. and Nitonye, S. (2016) Design Modeling and Performance Optimization of a Marine Boiler. World Journal of Engineering Research and Technology, 2, 44-58.

2. Liptak, B. (2006) Process Control and Optimization. 4th Edition, Taylor and Francis Group, United States of America.
3. Boisson, P. (1999) Safety at Sea: Policies, Regulations & International Law. Bureau Veritas, Paris.
4. Adumene, S. and Nitonye, S. (2016) Assessment of Site Parameters and Heat Recovery Characteristics on Combined Cycle Performance in an Equatorial Environment. World Journal of Engineering and Technology, 4, 313-324. https://doi.org/10.4236/wjet.2016.42032
5. Raunek (2009) Marine Insight. http://www.marineinsight.com
6. Adamkiewicz, A. and Janusz, F. (2013) Application of Risk Analysis in Maintenance of Ship Power System Elements. Scientific Journals of the Maritime University of Szczecin, 36, 5-12.
7. Adamkiewicz, A. and Burnos, A (2009) Influence of Maintenance Strategies on the Reliability of Gas Turbine in Power Systems of Floating Production, Storage and offloading Units. 28th International Scientific Conference on DIAGO, Ostrava, 2009, 5 – 13.
8. Nitonye, S., Adumene, S. and Howells, U.U. (2017) Numerical Design and Performance Analysis of a Tug Boat Propulsion System. Journal of Power and Energy Engineering, 5, 80-98. https://doi.org/10.4236/jpee.2017.511007
9. Samson, N. (2017) Numerical Analysis for the Design of the Fuel System of a Sea Going Tug Boat in the Niger Delta. World Journal of Engineering Research and Technology, 3, 161-177.
10. Taylor, D.A. (1996) Introduction of Marine Engineering. Butterworth-Heinemann, Oxford.
11. Ogbonnaya, E.A., Orji, J.C., Ugwu, H.U., Poku R., and Samson, N. (2014) Condition Monitoring and Fault Diagnosis of a Steam Boiler Feed Pump. International research journal in Engineering, Science and Technology, 11, 13-21.
12. NFPA (2011) NEPA 85: Boiler and Combustion System Harzard Code. National Fire Protection Association, Quincy.
13. Nwaoha, T.C., Andrew, J. and Adumene, S. (2015) Incorporation of Novel Model in Failure Analysis of Propeller Operations of Sea Going Vessels. Ships and Offshore Structures, 12, 9-18. https://doi.org/10.1080/17445302.2015.1099226
14. Nwaoha T.C., Yang, Z., Wang, J. and Bonsall, S. (2010) Application of Genetic Algorithm to Risk-Based Maintenance Operations of Liquefied Natural Gas Carrier Systems. Proceedings of the Institution of Mechanical Engineers, Part E: Journal of Process Mechanical Engineering, 225, 40-52. https://doi.org/10.1243/09544089JPME336
15. Norman, B.F. (1987) Reliability Engineering for Electronic Design. Marcel Dekker Incorporated, New York.
16. Pillay, A. and Wang, J. (2003) Technology and Safety of Marine Systems. Vol. 7, Elsevier Ocean Engineering Book Series, Oxford.
17. Desmond, N.D.H. and Gregory, B.B. (2004) Risk and Uncertainty in Dam Safety. Thomas Telford Ltd., London.
18. Wang, J. and Trbojevic, V.M. (2007) Design for Safety of Marine and Offshore Systems. Institute of Marine Engineering, Science and Technology, London.
19. Bartlett, L.M. (2000) Variable Ordering Heuristics for Binary Decision Diagrams. Doctoral Thesis, Loughborough University, Loughborough.

20. Bartlett, L.M. and Andrews, J.D. (2002) Choosing an Ordering Heuristic for the Fault Tree to Binary Decision Diagram Conversion Using Neural Networks. IEEE Transactions on Reliability, 51, 344-349. https://doi.org/10.1109/TR.2002.802892

21. Amari, S., Dill, G. and Howald, E. (2003) A New Approach to Solve Dynamic Fault Trees. Proceedings of the Annual Reliability and Maintainability Symposium, Tampa, 27-30 January 2003, 374-379. https://doi.org/10.1109/RAMS.2003.1182018

22. Takehis, K. (2006) A Simple Method to Derive Minimal Cut Sets for a Non-Coherent Fault Tree. International Journal of Automation and Computing, 3, 151-156. https://doi.org/10.1007/s11633-006-0151-4

23. Harvey, J.F. (2008) Enhancement of Operational Safety of Engine Room Machinery through (CBT).

24. Frederick (2013) High Pressure Boiler. American Technical Publishers, Orland Park.

25. SINTEF (2002) Offshore Reliability Data (OREDA). A Handbook Prepared by SINTEF Technology and Society on the Behalf of the OREDA Project, 4th Edition.

26. Hyo, K., Jae-Sun, K., Youngsoo, K. and Theofanius, G.T. (2005) Risk Assessment of Membrane Type LNG Storage Tanks in Korea-Based on Fault Tree Analysis. Korean Journal of Chemical Engineering, 22, 1-8. https://doi.org/10.1007/BF02701454

27. Risknology (2006) Failure Rate Data to the Independent Risk Analysis of the Cabrillo Port LNG Deepwater Port. A Report Prepared for U.S. Coast Guard Agency.

CHAPTER

7

Possible Efficiency Increasing of Ship Propulsion and Marine Power Plant with the System Combined of Marine Diesel Engine, Gas Turbine and Steam Turbine

Marek Dzida

Gdansk University of Technology Poland

1. INTRODUCTION

For years there has been, and still is, a tendency in the national economy to increase the efficiency of both the marine and inland propulsion systems. It is driven by economic motivations (rapid increase of fuel prices) and ecological aspects (the lower the fuel consumption, the lower the emission of noxious substances to the atmosphere). New design solutions are searched to increase the efficiency of the propulsion system via linking Diesel engines with other heat engines, such as gas and steam turbines. The combined systems implemented in marine propulsion systems in recent years are based mainly on gas and steam turbines (MAN, 2010). These systems can reach the efficiency exceeding 60% in inland applications. The first marine system of this type was applied on the passenger liner "Millenium". However, this is the only high-efficiency marine application of the combined propulsion system so far. Its disadvantage is that the system needs more expensive fuel, the marine Diesel oil, while the overwhelming majority of the merchant ships are driven by low-speed engines fed with relatively cheap heavy fuel oil. It seems that the above tendency will continue in the world's merchant navy for the next couple of years.

The compression-ignition engine (Diesel engine) is still most frequently used as the main engine in marine applications. It burns the cheapest heavy fuel oil and reveals the highest efficiency of all heat engines. The exhaust gas leaving the Diesel engine contains huge energy which can be utilised in another device (engine), thus increasing the efficiency of the entire system and reducing the emission of noxious substances to the atmosphere.

A possible solution here can be a system combined of a piston internal combustion engine and the gas and steam turbine circuit that utilises the heat contained in the exhaust gas from the Diesel engine. The leading engine in this system is the piston internal combustion engine. It seems that now, when fast container ships with transporting capacity of 8-12 thousand TU are entering into service, the propulsion engines require very large power, exceeding 50-80 MW. On the other hand, increasing prices of fuel and restrictive ecological limits concerning the emission of NO_x and CO_2 to the atmosphere provoke the search for new solutions which will increase the efficiency of the propulsion and reduce the emission of gases to the atmosphere.

The ship main engines will be large low-speed piston engines that burn heavy fuel oil. At present, the efficiency of these engines nears 45 – 50%. For such a large power output ranges, the exhaust gas leaving the engine contains huge amount of heat available for further utilisation.

The proposed combined system consisting of a piston internal combustion engine, a gas turbine and a steam turbine can also be used for engines of lower power, ranging between 400 ÷ 900 kW. For those power ranges a use of low-boiling media of organic-based refrigerant type instead of water (steam) in the steam cycle seems to be a reasonable solution. Piston internal combustion engines of this power range are used on coasting vessels, or in the inland water transport, for instance for driving cargo barges. On the other hand in inland applications the power blocks fired with solid, liquid, or gas fuels are in almost 100% the systems with steam or gas turbines.

In the Central Europe, Poland for instance, the basic fuel in power engineering is coal. Conventional electric power plants have the efficiency of an order of 38-42%, and emit large volumes of CO_2, NO_x and/ or SO_x. In order to decrease the amount of noxious substances emitted to the atmosphere and reduce the cost of production of the electric energy, combined systems are in use - consisting of gas turbines with a steam turbine circuit.

On the other hand, the combined turbine power plants can be complemented by electric power plants with a Diesel engine as the main propulsion. The exhaust gas leaving the engine contains about 30 – 40% of the heat delivered to the engine in the fuel. Using the heat from the exhaust gas in the gas and steam turbine circuit will increase the efficiency of the entire combined system. For large powers of piston internal combustion engines, the additional gas and steam turbine circuit is a source of measurable economic savings in electric energy production. Moreover, in large-power piston internal combustion engines we can additionally use the low-temperature waste heat, for instance for heating the communal water (Dzida, 2009). In the seaside areas with no large electric power plants, a mobile power plant situated on a platform close to the coast reveals additional advantages:

- increasing production of electric energy in the so-called distributed system,
- diversification of primary energy sources which decreases the consumption of coal in favour of liquid fuels,
- possible combustion of residual heavy fuels from nearby oil refineries,
- reducing large-distance transport of solid fuels, the absence of slag and cinders,

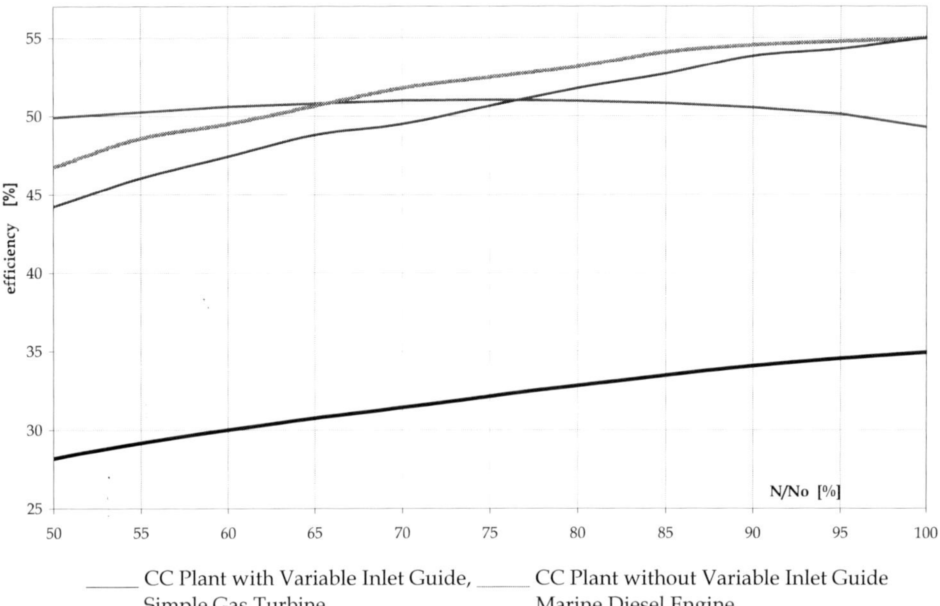

Figure 1. *Part - load Efficiency of a Combined - Cycle Plant (GT&ST), Simple Gas Turbine and Marine Diesel Engine*

- reducing the emission of CO_2 and NO_x due to the increased system efficiency,
- shorter time of plant erection compared to that of a conventional power plant, and possibility of opening it in stages: first with the Diesel engine alone, and then complementing it, during plant operation, with a combined steam/gas turbine system,
- no problems with the water cooling the condenser, small effect on the environment in water balance aspects,
- mobility of a combined power plant erected on the marine platform.

2. CONCEPT OF A COMBINED SYSTEM

Combined propulsion systems are used in marine engineering mostly in fast specialpurpose ships and in the Navy, as the systems being a combination of a Diesel engine and gas turbines (CODAG, CODOG) or solely gas turbines (COGOG, COGAG). The propulsion system of the passenger liner "Millenium" uses a COGES-type system which improved the efficiency and operating abilities of the ship. The system consists of a gas turbine and a steam turbine which drive an electric current generator, while the propeller screws are driven by electric motors. In this system the steam turbine circuit is supplied with the steam generated in the waste heat boiler supplied with the exhaust gas from the gas turbines.

Combined systems used in inland power blocks base on a gas turbine as the main unit and a steam turbine that utilises the steam produced in a waste heat boiler using the heat recovered from the gas turbine exhaust gas. All this provides

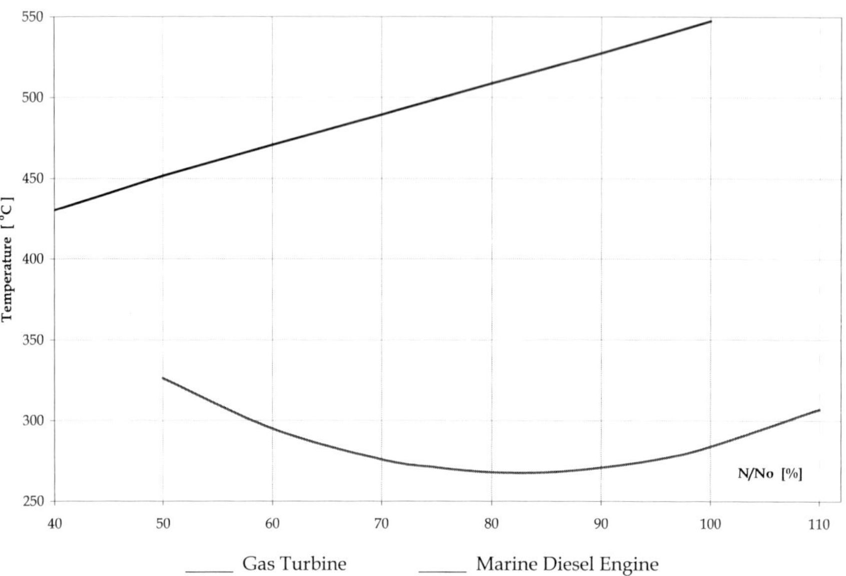

Figure 2. *Temperatures of the exhaust gas from the Diesel engine and the gas turbine as function of power plant load*

opportunities for reaching high efficiency of the combined block. The exhaust gas leaving a marine low-speed Diesel engine contains smaller amount of heat, of an order of 30 – 40% of the energy delivered to the engine.

Figure 1 shows sample efficiency curves of the combined gas turbine/steam turbine systems as functions of power plant load, compared to the gas turbine operating in a simple open circuit and the marine low-speed Diesel engine.

The efficiency curves in Fig. 1 show that the combined cycle gas turbine/steam turbine system has the highest efficiency for maximal loads (maximal efficiency levels for these circuits reach as much as 60%). Gas turbines operating in the simple open circuit have the lowest efficiency (average values of 33÷35%, and maximal values reaching 40%). Low-speed Diesel engines have the efficiency of an order of 47 ÷ 50%. It is also noticeable that the Diesel engine curve is relatively flat. This is of special importance in case of marine propulsion systems which operate at heavily changing loads. For the combined cycle gas turbine/steam turbine systems and the gas turbines operating in the simple open circuit the relative efficiency decrease $\Delta\eta / \eta$ is equal to 15 ÷ 20% when the load decreases from 100% to 50%. For the low-speed Diesel engine these numbers are equal to 1 ÷ 2%. This property of the Diesel engine, along with the ability to utilise additional heat contained in its exhaust gas, makes the engine the most applicable in marine propulsion systems operating in heavily changing load conditions. The amount of heat contained in the exhaust gas from the gas turbine is approximately equal to 60 ÷ 65%, i.e. more than in piston engines, which results from lower exit temperature and less intensive flow of the exit gas leaving the Diesel engine, Figs. 2 and 3.

The exit temperatures of the exhaust gas from the gas turbines range between 450 ÷ 600°C, on average, while those from the low-speed Diesel engines are of an

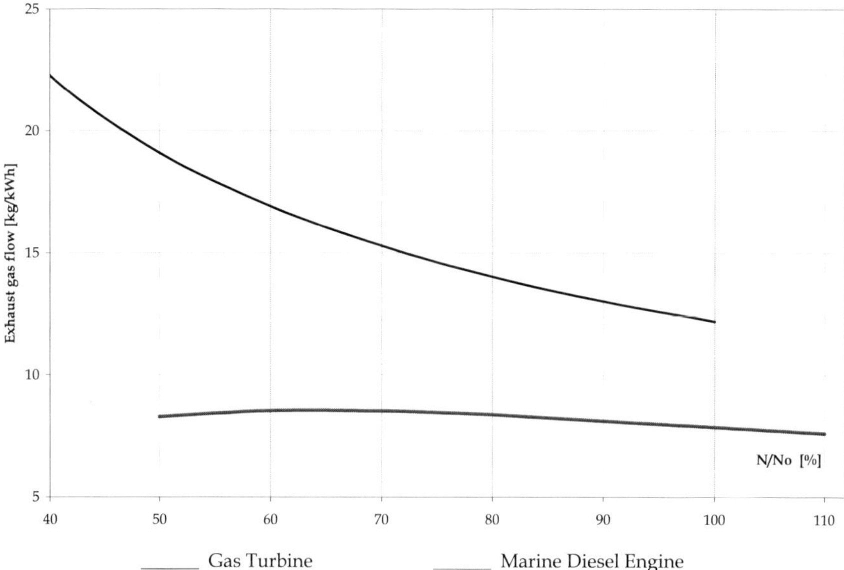

Figure 3. *Related exhaust gas mass flow rate as a function of power plant load*

order of 220 ÷ 300°C. In the gas turbines, decreasing the load remarkably decreases the temperature of the exhaust gas, while in the Diesel engine these changes are much smaller, and the temperature initially decreases and then starts to increase for low loads.

This property of the steam turbine circuit in the combined system with the Diesel engine for partial loads makes it possible to keep the live steam temperature at a constant level within a wide range of load. The related exhaust gas mass flow rate $m_g/N[kg/kWh]$ changes only by about 5% in the Diesel engine when the load changes from 100% to 50%, while in the gas turbine this parameter changes by about 55% for the same load change, Fig. 3.

The combined propulsion system with the low-speed piston internal combustion engine used as the main engine and making use of the heat from the engine exhaust gas is shown in Fig. 4, (Dzida, 2009; Dzida & Mucharski, 2009; Dzida et al., 2009).

The exhaust gas flows leaving individual main engine cylinders are collected in the exhaust manifold and passed to the constant-pressure turbocharger. Due to high turbocharger efficiency ranges (MAN, 2010; Schrott, 1995), the scavenge air can be compressed using the energy contained only in part of the exhaust gas flow. The remaining part of the exhaust gas flow can be expanded in an additional gas turbine, the so-called power turbine, which additionally drives, via a gear, the propeller screw or the electric current generator.

The exhaust gas from the turbocharger and the power turbine flows to the waste heat boiler installed in the main engine exhaust gas path, before the silencer. The waste heat boiler produces the steam used both for driving the steam turbine that passes its energy to the propeller screw, and for covering all-ship needs.

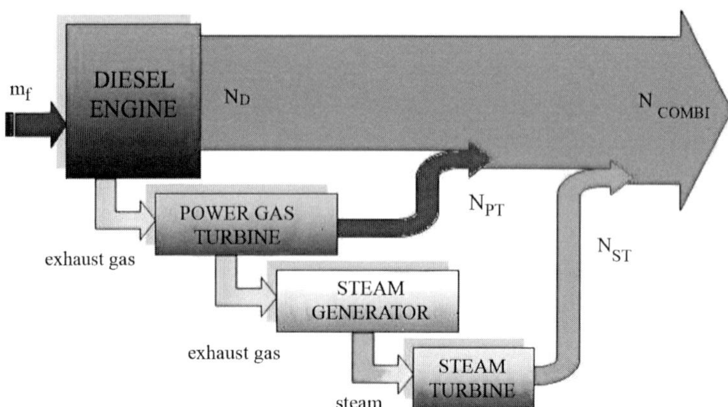

Figure 4. *Concept of the combined propulsion system*

In the marine low-speed Diesel engines, another portion of energy that can be used along with the exhaust gas energy is a huge amount of so-called waste heat of relatively low temperature. In the low-speed engines the waste heat comprises the following components (with their proportions to the heat delivered to the engine in fuel):

- heat in the scavenge air cooler (17-20%), of an approximate temperature of about 200°C,
- heat in the lubricating oil cooler (3-5%), of an approximate temperature of about 50%.
- heat in the jacket water cooler (5-6%), of the temperature of an order of 100°C.

This shows that the amount of the waste heat that remains for our disposal is equal to about 25 – 30% of the heat delivered in fuel. Part of this heat can be used in the combined circuit with the Diesel engine.

2.1. Energy evaluation of the combined propulsion system

The adopted concept of the combined ship propulsion system requires energy evaluation, Fig. 4. Formulas defining the system efficiency are derived on the basis of the adopted scheme.

The power of the combined propulsion system is determined by summing up individual powers of system components (the main engine, the power gas turbine, and the steam turbine):

$$N_{combi} = N_D + N_{PT} + N_{ST} \tag{1}$$

hence the efficiency of the combined system is:

$$\eta_{\text{combi}} = \frac{N_{\text{combi}}}{m_{fD} \cdot Wu} = \eta_D \cdot \left(1 + \frac{N_{PT}}{N_D} + \frac{N_{ST}}{N_D}\right) \tag{2}$$

and the specific fuel consumption is:

$$b_{\text{ecombi}} = b_{eD} \cdot \frac{1}{\left(1 + \frac{N_{PT}}{N_D} + \frac{N_{ST}}{N_D}\right)} [g/kWh] \tag{3}$$

where η_D, b_{CD} – is the efficiency and specific fuel consumption of the main engine.

Relations (2) and (3) show that each additional power in the propulsion system increases the system efficiency and, consequently, decreases the fuel consumption. And the higher the additional power achieved from the utilisation of the heat in the exhaust gas leaving the main engine, the lower the specific fuel consumption. Therefore the maximal available power levels are to be achieved from both the power gas turbine and the steam turbine. The power of the steam turbine mainly depends on the live steam and condenser parameters.

2.2. Variants of the combined ship propulsion systems or marine power plants

For large powers of low-speed engines, the exhaust gas leaving the engine contains huge amount of heat available for further utilisation. Marine Diesel engines are always supercharged. Portions of the exhaust gas leaving individual cylinders are collected in the exhaust gas collector, where the exhaust gas pressure $p_{exh\_D} > p_{bar}$ is equalised. In standard solutions the constant-pressure turbocharger is supplied with the exhaust gas from the exhaust manifold to generate the flow of the scavenge air for supercharging the internal combustion engine.

Present-day designs of turbochargers used in piston engines do not need large amounts of exhaust gas, therefore it seems reasonable to use a power gas turbine complementing the operation of the steam turbine in those cases. Here, two variants of power gas turbine supply with the exhaust gas are possible.

2.2.1. Parallel power gas turbine supply (variant A)

In this case part of the exhaust gas from the piston engine exhaust manifold supplies the Diesel engine turbocharger. The remaining part of the exhaust gas from the manifold is directed to the gas turbine, bearing the name of the power turbine (PT). The power turbine drives, via the reduction gear, the propeller screw or the electric current generator, thus additionally increasing the power of the entire system. Figure 5 shows a concept of this propulsion system, referred to as parallel power turbine supply. After the expansion in the turbocharger and the power turbine, the exhaust gas flowing from these two turbines is directed to the waste heat boiler in the steam circuit.

In the proposed solution, at low load ranges the amount of the exhaust gas from the main engine is not sufficient to additionally supply the power turbine. In such case a control valve closes the exhaust gas flow to the power turbine, Figure 5. The operation of this valve is controlled by the control system using two signals: the scavenge air pressure signal, and the signal of the propeller shaft angular speed or torque. The waste heat boiler produces the steam which is then used both in the steam turbine and, in case of marine application, to cover the all-ship needs. This system allows for independent operation of the Diesel engine, with the steam

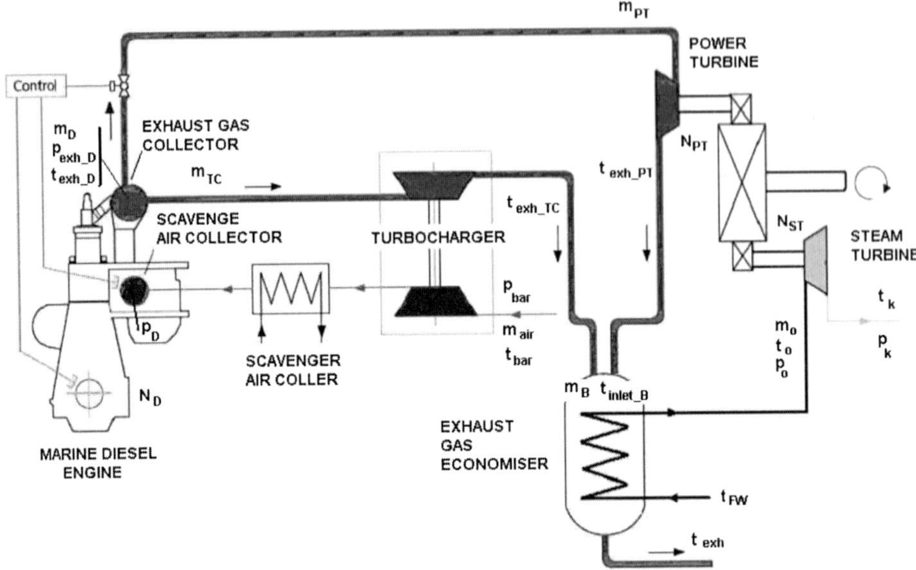

Figure 5. *Combined system with the Diesel main engine, the power turbine supplied in parallel, and the steam turbine (variant A)*

turbine or the power turbine switched off. The control system makes it possible to switch off the power turbine thus increasing the power of the turbocharger at partial load, and, on the other hand, direct part of the Diesel engine exhaust gas to supply the power turbine at large load.

Power turbine calculations are based on the Diesel engine parameters, i.e. the temperature of the exhaust gas in the exhaust gas collector, which in turn depends on the engine load and air parameters at the engine inlet. Marine engine producers most often deliver the data on two reference points for the atmospheric air (the ambient reference conditions):

ISO Conditions

Tropical Conditions

Ambient air temperature [°C] 25 45

Barometric pressure [bar] 1

2.2.2. Series power gas turbine supply (variant B)

In this variant the exhaust gas from the exhaust manifold supplies first the piston engine turbocharger and then the power turbine, Fig.6.

After leaving the exhaust manifold, the exhaust gas expands in the turbochar ger to the higher pressure than the atmospheric pressure, which leaves part of the exhaust gas enthalpy drop for utilisation in the power turbine. The exhaust gas leaving the power turbine passes its heat to the steam in the waste heat boiler, thus producing additional power in the steam turbine circuit.

Also in this combined system, the installed control valve makes it possible to switch off the power turbine at partial piston engine loads, thus increasing the

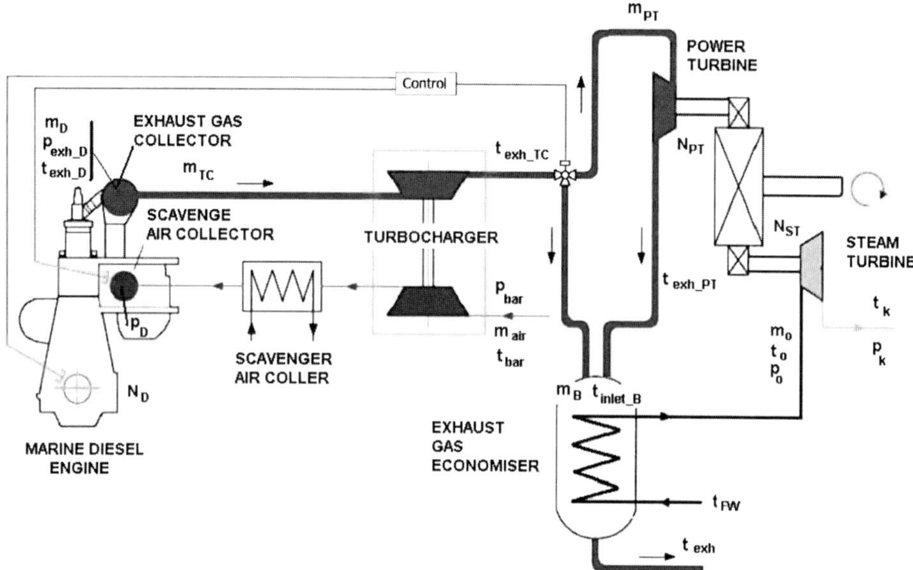

Figure 6. *Combined system with the Diesel main engine, the power turbine supplied in series, and the steam turbine (variant B)*

power of the turbocharger by expanding the exhaust gas to lower pressure, Fig. 6. Unlike the parallel supply variant, here the entire mass of the exhaust gas from the piston engine manifold flows through the turbocharger. The exhaust gas pressure at the turbocharger outlet is higher than in variant A.

3. POWER TURBINE IN THE COMBINED SYSTEM

Calculating the power turbine in the combined system depends on the selected variant of power turbine supply. Usually, piston engine producers do not deliver the exhaust gas temperature in the exhaust manifold (which is equal to the exhaust gas temperature at turbocharger turbine inlet). Instead, they give the exhaust gas temperature at turbocharger turbine outlet ($t_{\text{exh\_D}}$). The temperatures of the exhau st gas in the Diesel engine exhaust gas collector are calculated from the turbine power balance, according to the following formula:

$$t_{\text{exh\_D}} = \frac{t_{\text{exh\_TC}} + 273,15}{1 - \eta_T \cdot \left(1 - \frac{1}{\pi_T^{\frac{kg-1}{kg}}}\right)} - 273,15\,[^\circ\text{C}] \tag{4}$$

This formula needs the data on turbocharger turbine efficiency changes for partial loads. These data can be obtained from the producer of the turbocharger (as they are rarely made public), Fig. 7, or calculated based on the relation used in steam turbine stage calculations:

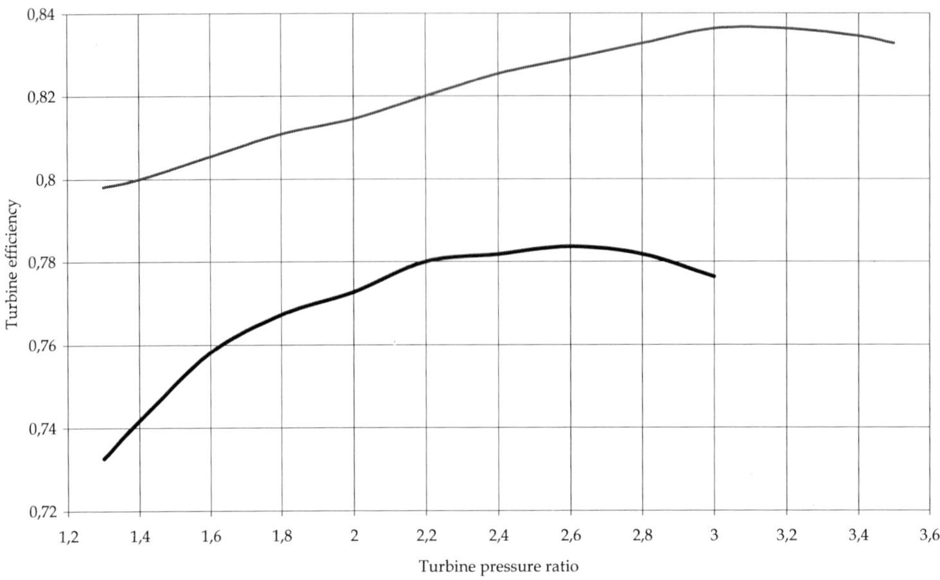

Figure 7. *Turbocharger turbine efficiency as a function of scavenge air pressure, acc. to (Schrott, 1995)*

$$\overline{\eta_T} \equiv \frac{\eta_T}{\eta_{T_o}} = 2 \cdot \bar{v} - \bar{v}^2 \tag{5}$$

where v - related turbine speed indicator, η_{To} – maximal turbine efficiency and the corresponding speed indicator.

The turbine speed indicator is defined as:

$$v = \frac{u}{c_s} = \sqrt{\frac{u^2}{2 \cdot H_T}} \tag{6}$$

where u- circumferential velocity on the turbine stage pitch diameter, H_T– enthalpy drop in the turbine.

The calculations make use of static characteristics of the turbocharger compressor, with the marked line of cooperation with the Diesel engine, Fig.8.

Figure 9 shows the turbocharger efficiency curves calculated from the relation:

$$\eta_{\mathrm{TC}} = \eta_{\mathrm{T}} \cdot \eta_{\mathrm{C}} \cdot \eta_m \tag{7}$$

where η_T - the turbocharger turbine efficiency is calculated from relation (5), while the compressor efficiency η_C is calculated from the line of Diesel engine/compress or cooperation, η_m - mechanical efficiency of the turbocharger, Fig. 8. In the same figure a comparison is made between the calculated turbocharger turbine efficiency with the producer's data as a function of the Diesel engine scavenge pressure. The differences between these curves do not exceed 1,5%.

For the presently available turbocharger efficiency ranges, the amount of the exhaust gas needed for driving the turbocharger turbine is smaller than the entire mass flow rate of the exhaust gas leaving the Diesel engine. Fig. 10 shows sample curves of exhaust gas temperature changes in the engine manifold (calculated using the relation (4)) and the exhaust gas temperature at the turbocharger outlet (according to the data delivered by the producer) as functions of engine load, when the standard internal combustion engine exhaust gas is expanded to the barometric pressure. The figure also shows the Diesel engine exhaust gas flow rate related to the scavenge air flow rate, as a function of the engine load. This high efficiency of the turbocharger provides opportunities for installing a power gas turbine connected in parallel with the turbocharger (variant A).

Δ calculated efficiency

The turbocharger power balance indicates that in the power gas turbine we can utilise between 10 and 24% of the flow rate of the exhaust gas leaving the exhaust manifold of the piston engine. The power gas turbine can be switched on when the main engine power output exceeds 60%. For lower power outputs the entire exhaust gas flow leaving the Diesel engine is to be used for driving the turbocharger.

In variant *B* of the combined system with the power turbine, the turbocharger is connected in series with the power gas turbine. Here, the entire amount of the exhaust gas flows through the turbocharger turbine. Due to the excess of the power needed for driving the turbocharger, the final expansion pressure at turbocharger turbine output can be higher than the exhaust gas pressure at waste heat boiler inlet. In this case the expansion ratio in the turbocharger turbine is given by the relation:

$$\pi_T = \left[\frac{1}{1 - \frac{1}{\eta_{\mathrm{TC}}} \cdot \frac{m_a}{m_D} \cdot \frac{c_a}{c_g} \cdot \frac{t_a}{t_{exh_D D}} \cdot \left(\pi_{\mathrm{C}}^{\frac{\aleph_a - 1}{\aleph_a}} \right)} \right]^{\frac{\aleph_g}{s_g^{-1}}} \qquad (8)$$

where: π_{C} - compression ratio of the turbocharger compressor.

The exhaust gas temperature at turbocharger outlet is calculated from the formula:

$$t_{\mathrm{exh\_TC}} = \left(t_{\mathrm{exh\_D}} + 273,15 \right) \cdot \left[1 - \eta_T \left(1 - \frac{1}{\pi_T^{\frac{T-1}{kg}}} \right) \right] - 273,15\ [^{\circ}\mathrm{C}] \qquad (9)$$

Figure 11 shows sample curves of temperature, compression and expansion rate changes in the turbocharger for variant B : series power turbine supply.

This case provides opportunities for utilising the enthalpy drop of the expanding exhaust gas in the power turbine. The operation of the power turbine is possible when the Diesel engine power exceeds 60%.

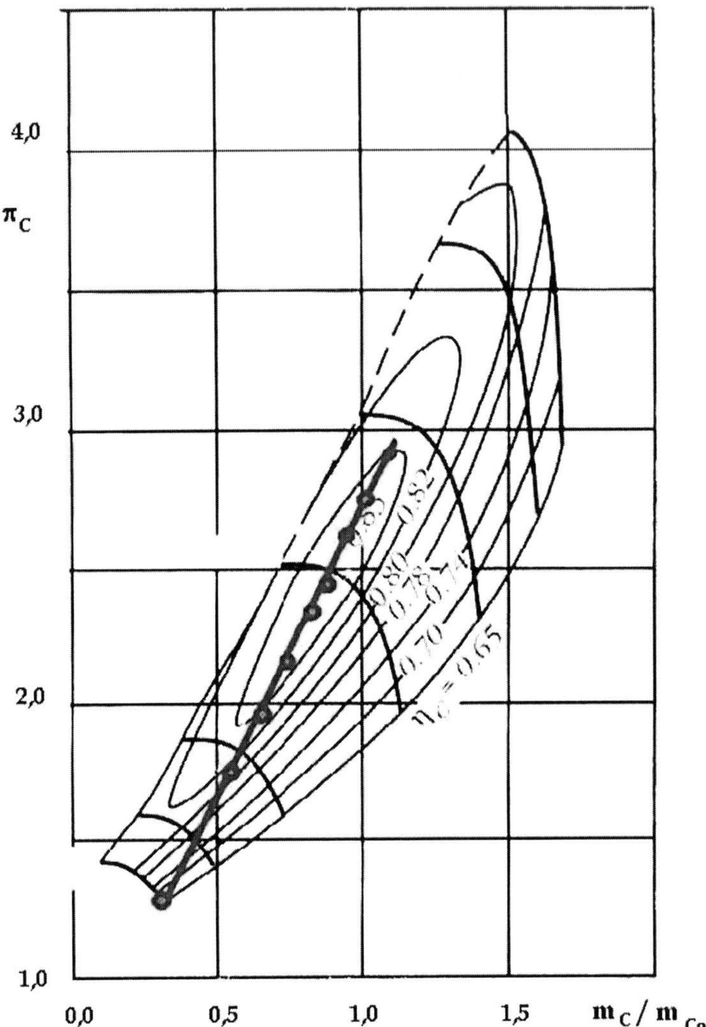

Figure 8. *Diesel engine cooperation line against turbocharger compressor characteristics*

3.1. Power turbine in parallel supply system (variant A)

The power turbine (Fig.5) is supplied with the exhaust gas from the exhaust manifold. The exhaust gas mass flow rate m_{PT} and temperature t_{exh_D} are identical as those at turbocharger outlet: the mass flow rate of the exhaust gas flowing through the power turbine results from the difference between the mass flow rate of the Diesel engine exhaust gas and of that expanding in the turbocharger:

$$m_{TD} = m_a \cdot (1 - \bar{m}) + m_{fD} \tag{10}$$

The mass flow rate of the exhaust gas needed by the turbocharger is calculated from the turbocharger power balance using the following formula:

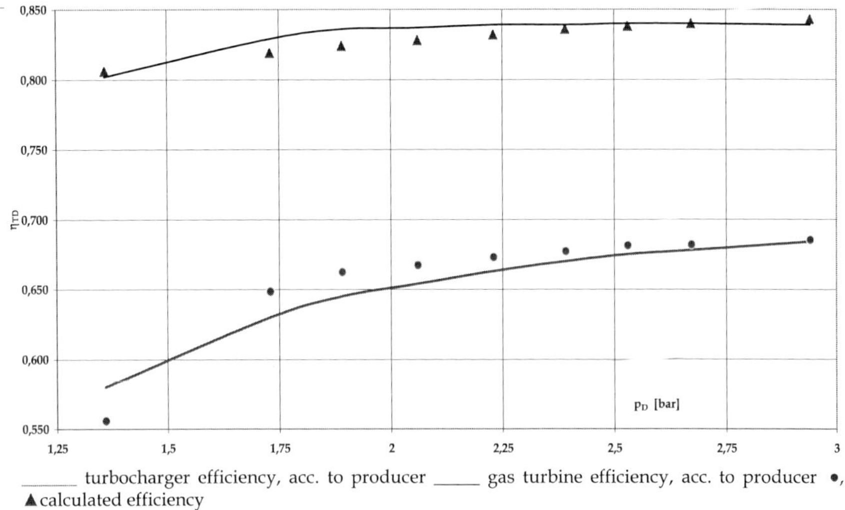

Figure 9. *Efficiency characteristics of the turbocharger and the turbocharger gas turbine as a function of scavenge air pressure*

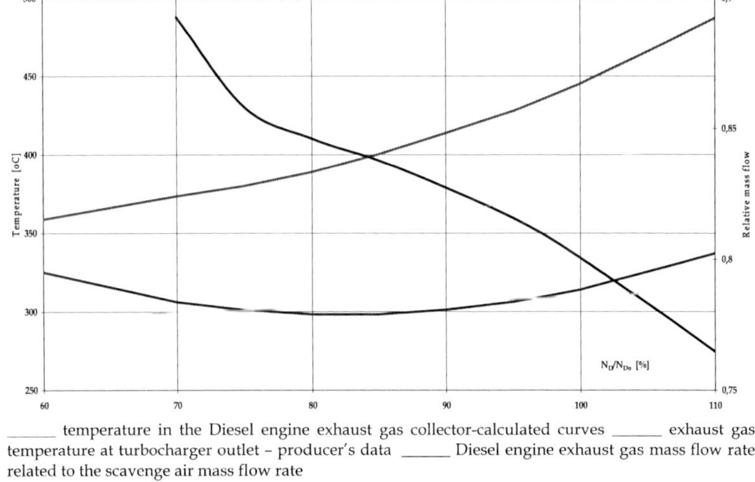

Figure 10. *Sample temperature characteristics of the turbocharger during gas expansion in the turbine to the atmospheric pressure and the related exhaust gas mass flow rates as functions of Diesel engine load*

$$1 - \frac{1}{\frac{\kappa g^{-1}}{kg}}$$

$$\bar{m} \equiv \frac{m_{TC}}{m_a} = \frac{\pi_T^{\frac{k_T}{k_g}}}{\pi_C^{\frac{k_k - 1}{k_a}} - 1} \cdot \frac{T_{\text{erh\_D}}}{T_a} \cdot \frac{c_g}{c_p} \cdot \eta_{TC} \tag{10.1}$$

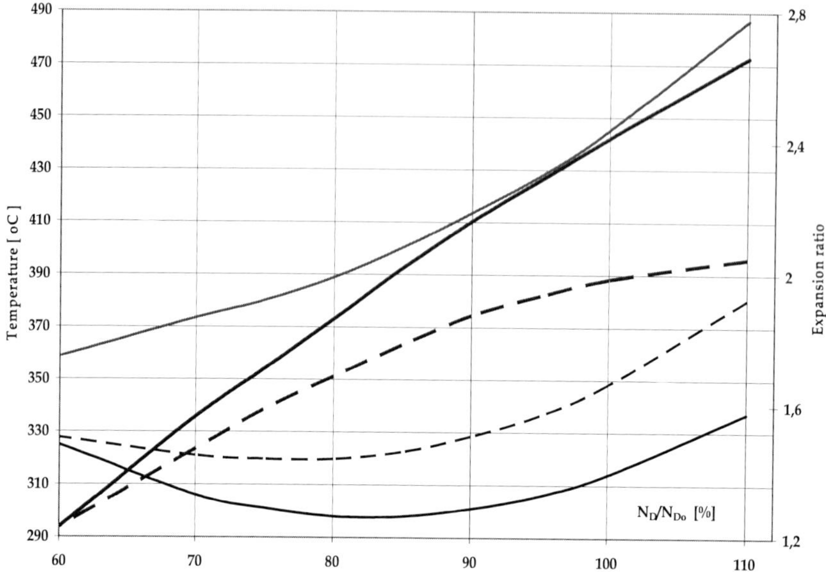

_____expansion ratio in the turbocharger turbine (standard arrangement - without power turbine) _ _ _ expansion ratio in the turbocharger turbine with power turbine ______exhaust gas temperature in the Diesel engine exhaust gas collector ____exhaust gas temperature at turbocharger outlet without power turbine _ _ _ exhaust gas temperature at turbocharger outlet with power turbine

Figure 11. *Changes of temperature and expansion ratio of the turbocharger in the combined system with series power turbine supply (variant B)*

The exhaust gas expanding in the power turbine has the inlet and outlet pressures identical to those of the exhaust gas flowing through the turbocharger. The power of the power turbine is given by the relation:

$$N_{PT} = \eta_m \cdot \eta_{PT} \cdot m_{PT} \cdot H_{PT} \tag{11}$$

where η_m - mechanical efficiency of the power turbine, H_{PT} - iso-entropic enthalpy drop in the power turbine.

The power turbine efficiency η_{PT} is assumed in the same way as for the turbocharger turbine, Fig. 9, or using the relation (5). In the shipbuilding, the gas turbines used in combined Diesel engine systems with power turbines are those adopted from turbochargers. The power turbine system calculations show that the exhaust gas temperature at the power turbine outlet is slightly higher than that at the turbocharger outlet, Fig.12. The increase of the main engine load results in the increase of both the exhaust gas temperature in the exhaust gas collector and the mass flow rate of the exhaust gas flowing through the power turbine. The increase in power of the combined system with additional power turbine ranges from about 2% for Diesel engine loads of an order of 70% up to over 8% for maximal loads, Fig.12.

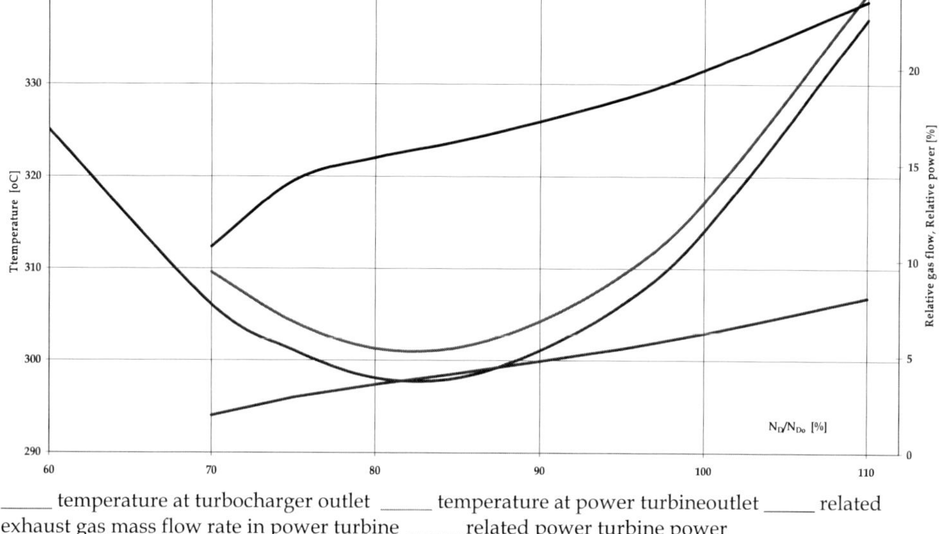

_____ temperature at turbocharger outlet _____ temperature at power turbineoutlet _____ related exhaust gas mass flow rate in power turbine _____ related power turbine power

Figure 12. *Parameters of parallel supplied power turbine as functions of the main engine load variant A (calculations for tropical conditions)*

When the Diesel engine power is lower than 60 – 70% of the nominal value the entire exhaust gas flow from the exhaust manifold is directed to the turbocharger drive. In this case the control system closes the valve controlling the exhaust gas flow to the power turbine, Fig. 5.

3.2. Power turbine in series supply system (variant B)

In this variant the power turbine is supplied with the full amount of the exhaust gas leaving the Diesel engine exhaust manifold. The power turbine is installed after the turbocharger. The exhaust gas pressure at the power turbine inlet depends on the pressure of the exhaust gas leaving the turbocharger turbine, Fig.11.

In this case the power of the power turbine is calculated as:

$$N_{PT} = \eta_{PT} \cdot m_D \cdot c_g \cdot t_{inl\_PT} \cdot \left(1 - \frac{1}{\pi_{PT}^{\frac{k-1}{kg}}}\right) \tag{12}$$

where t_{inl_PT} - exhaust gas temperature at the power turbine inlet, π_{PT} - expansion ratio in the power turbine, η_{PT}- power turbine efficiency. The power turbine efficiency is assumed in the same way as in variant A.

In formula (12) the exhaust gas temperature at the power turbine inlet is assum ed equal to that of the exhaust gas leaving the turbocharger, Fig. 13.

The exhaust gas temperature at the power turbine output is calculated from the formula:

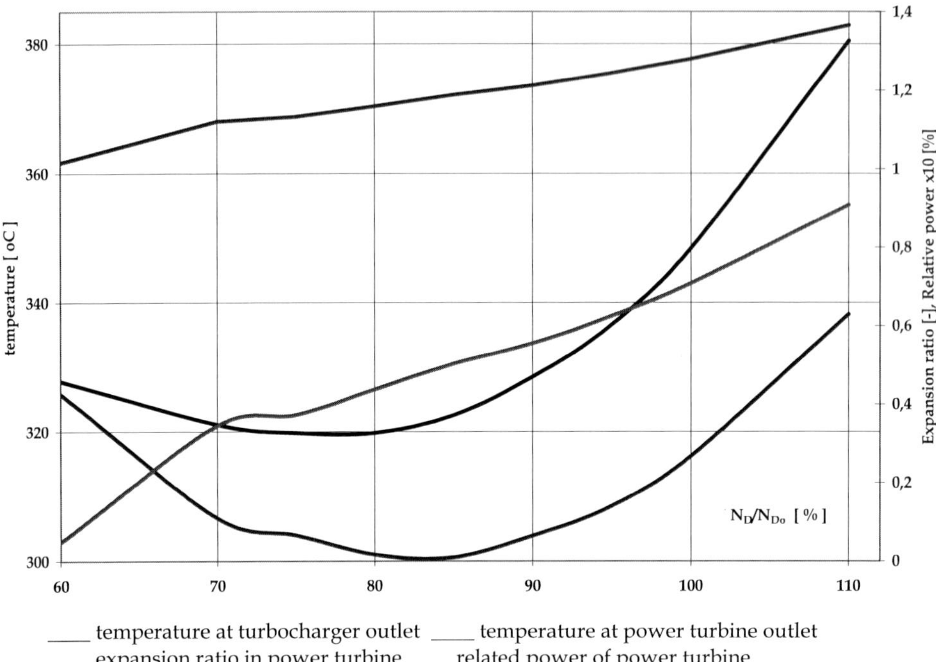

Figure 13. *Parameters of series supplied power turbine as functions of the main engine load variant B (calculations for tropical conditions)*

$$t_{exh\_PT} = \left(t_{inl\_PT} + 273{,}15\right) \cdot \left[1 - \eta_{PT}\left(1 - \frac{1}{\pi_{PT}^{\frac{\aleph_g - 1}{\aleph_g}}}\right)\right] - 273{,}15[^{\circ}C] \tag{13}$$

Figure 13 also shows the expansion ratio, the power of the power turbine, and the exhaust gas temperatures at the turbocharger and the power turbine outlets for partial engine loads. The power turbine in this variant increases the power of the combined system by 3% to 9% with respect to that of a standard engine. The turbine power increases with increasing Diesel engine load.

3.3. Comparing the two power turbine supply variants

The analysis of the two examined variants shows that the power of the combined system increases depending on the Diesel engine load. For both variants the power turbine can be used after exceeding about 65% of the Diesel engine power. The exhaust gas leaving the power turbine is directed to the waste heat boiler, where together with steam turbine it can additionally increase the overall power of the combined system.

In both cases the temperatures of the exhaust gas leaving the power turbine are comparable. The exhaust gas pressure at power turbine outlet depends on the losses generated when the gas flows through the waste heat boiler and outlet silencers. Following practical experience, the exhaust gas back pressure is assumed higher than the barometric pressure by 300 mmWC, i.e. about 3%. Taking into account powers of the power turbines for the above variants, Fig. 14, it shows that for the same Diesel engine parameters the series supply of the power turbine results in higher turbine power. For lower loads, the power of the series supplied power turbine increases, compared to the parallel supply variant.

4. STEAM TURBINE CIRCUIT

The combined system makes use of the waste heat from the Diesel engine. In modern Diesel engines the temperatures of the waste heat are at the advantageous levels for the steam turbine circuit. This circuit makes use of water that can be utilised in a low-temperature process. Adding the steam circuit to the combined Diesel engine/power gas turbine system provides good opportunities for increasing the power of the combined system, and consequently, also the system efficiency, see formula (2).

In the examined combined system the exhaust gas leaving the turbocharger and the power turbine (variant A, Fig. 5) or only the power turbine (variant B, Fig. 6) flows to the waste heat boiler where it is used for producing superheated steam for driving the steam turbine.

The mass flow rate of the exhaust gas reaching the waste heat boiler is equal to that leaving the Diesel engine exhaust gas collector. The exhaust gas temperature at waste heat boiler inlet depends on the adopted solution of power turbine supply. For variant A with parallel supply it is calculated from the balance of mixing of the gases leaving the turbocharger and the power turbine:

$$t_{inl\_B} = \frac{m_{TC} \cdot i_{ext\_TC} + m_{PT} \cdot i_{ext\_PT}}{m_D \cdot c_8} - 273{,}15 \, [^\circ C] \tag{14}$$

while for the series power turbine supply (variant B) it is assumed equal to that at the power turbine outlet, formula (13).

In combined steam turbine systems for small power ranges and low live steam temperatures the single pressure systems are used, Fig. 15, (Kehlhofer, 1991).

Such system consists of a single-pressure waste heat boiler, a condensing steam turbine, a water-cooled condenser, and a single stage feed water preheater in the deaerator.

The main disadvantage of the systems of this type is poor utilisation of the heat contained in the exhaust gas (the waste heat energy). The steam superheater is relatively large, as the entire mass of the steam produced by the boiler flows through it. However, costs of this steam system are the lowest, as poor utilisation of the exhaust gas energy results in high temperature of the exhaust gas leaving the boiler. The deaerator is supplied with the steam extracted from the steam turbine.

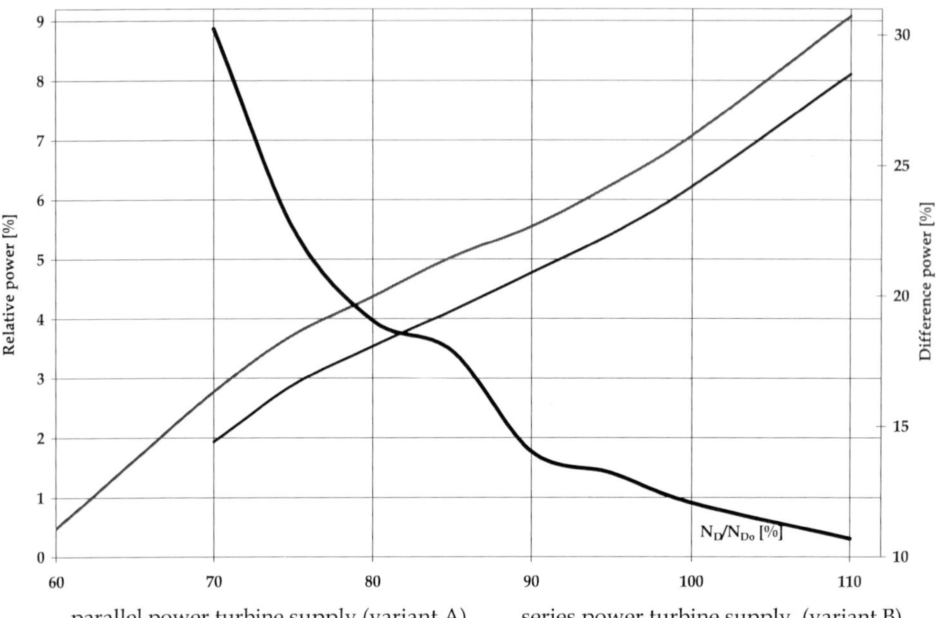

____ parallel power turbine supply (variant A) ____ series power turbine supply (variant B)

Figure 14. *Powers of the power turbine as functions of main engine load*

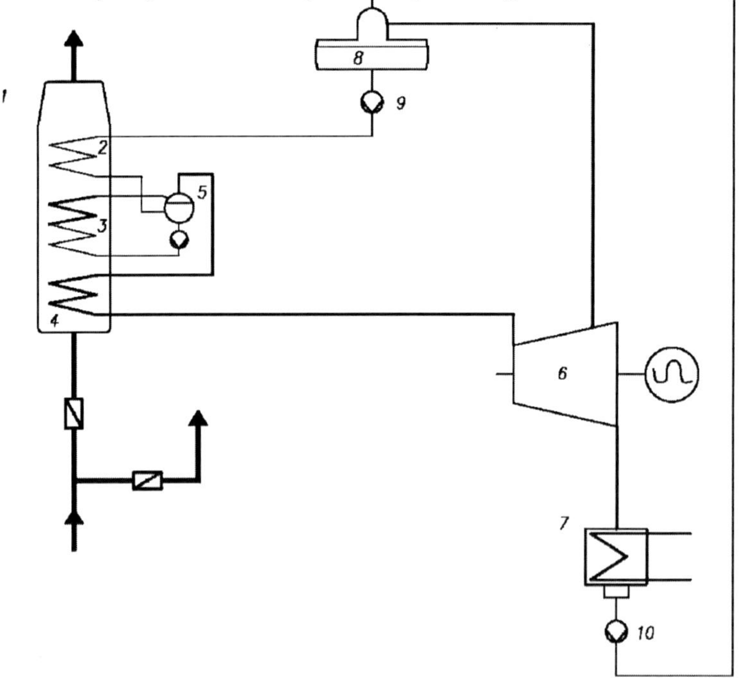

Figure 15. *Flow Diagram of the Single Pressure System*

The application of the single pressure system does not secure optimal utilisation of the exhaust gas energy.

Those steam turbine systems frequently make use of an additional low-press ure evaporator, Fig. 16, which leads not only to more intensive utilisation of the waste heat contained in the exhaust gas, but also to better thermodynamic use of the low-pressure steam.

In this solution the high pressure superheater is relatively small, compared to the single pressure boiler. The deaerator is heated with the saturated steam from the low-pressure evaporator. The power of the main high-pressure feeding pump is also smaller. The excess steam from the low-pressure evaporator can be used for supplying the low-pressure part of the steam turbine, thus increasing its power, or, alternatively, for covering all-ship needs. Figure 16 shows possible use of the temperature waste heat from the scavenge air cooler, the lubricating oil cooler, and from the jacket water cooler in the low-pressure water preheater.

The additional low-pressure exchanger in the steam circuit, Fig. 16, makes it possible to increase the temperature of the water in the deaerator. Higher water temperature is required due to the presence of sulphur in the fuel (water dew-point in the exhaust gas) - it is favourable for systems fed with a high sulphur content fuel. If the temperature of the feedwater is low when the system is fed with fuel without sulphur, the heat exchanger 14 in Fig. 16 is not necessary and the waste heat from the coolers can be used in the deaerator. For a low feedwater temperature the deaerator works at the pressure below atmospheric (under the vacuum).

4.1. Limits for steam circuit parameters

The limits for the values of the steam circuit parameters result from strength and technical requirements concerning the durability of particular system components, but also from design and economic restrictions. The difference between the exhaust gas temperature and the live steam temperature, Δt, for waste heat boilers used in shipbuilding is assumed as $\Delta t = 10\text{-}15C$, according to (MAN, 1985; Kehlhofer, 1991). The "pitch point" value recommended by MAN B&W (MAN, 1985) for marine boilers is $\delta t = 8 - 12°C$. The limiting dryness factor x of the steam downstream of the steam turbine is assumed as $x_{limit} = 0,86 - 0,88$. For marine condensers cooled with sea water, MAN recommends the condenser pressure $p_K = 0,065$ bar. This pressure depends on the B&W (MAN, 1985) temperature of the cooling medium in the condenser. Figure 17 shows the dependence of the condenser pressure on the cooling medium temperature. The temperature of the boiler feed water is of high importance for the life time of the feed water heater in the boiler. The value of this temperature is connected with a so-called exhaust gas dew-point temperature. Below this temperature the water condensates on heater tubes and reacts with the sulphur trioxide SO_3 producing the sulphuric acid, which is the source of low-temperature corrosion. That is why boiler producers give minimal feed water temperatures below which boiler operation is highly not recommended. The dew-point temperature is connected with the content of sulphur in the fuel and depends on the excess air coefficient in the piston engine. Figure 18

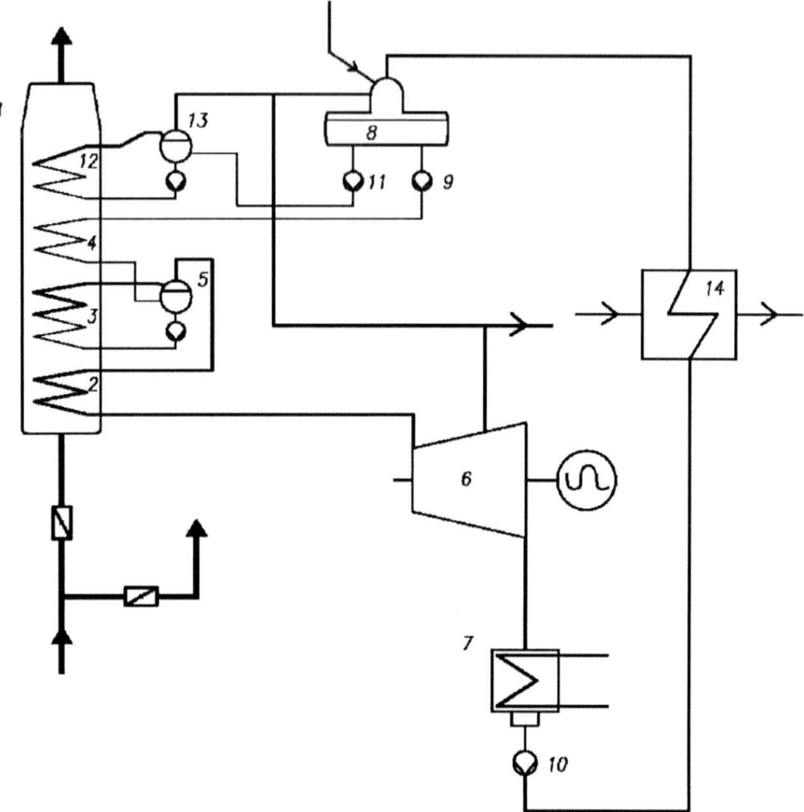

Figure 16. *Flow Diagram for a Two - Pressure System*

shows the dew-point temperature as the function of: sulphur content in the fuel, SO_2 conversion to SO_3, and the excess air coefficient in the engine. In inland power installations burning fuels with sulphur content higher than 2%, the recommended level of feed water temperature is $t_{FW} > 140 - 145°C$ (Kehlhofer, 1991).

In marine propulsion (MAN, 1985) recommends that the feed water temperature should not be lower than 120°C when the sulphur content is higher than 2%. This is justified by the fact that the outer surface of the heater tubes on the exhaust gas side has the temperature higher by 8 – 15°C than the feed water temperature, and that the materials used in those heaters reveal enhanced resistance to acid corrosion.

The exhaust gas temperature at the boiler outlet is assumed higher by 15 – 20°C than the feed water temperature, i.e. $t_{exh} > t_{FW} + (15 - 20°C)$.

Each ship burning heavy fuel in its power plant uses the mass flow rate mss of the saturated steam taken from the waste heat boiler for fuel pre-heating and all-ship purposes. According to the recommendations (MAN, 1985) the pressure of the steam used for these purposes should range between $p_{SS} = 7 - 9$ bar. This pressure is also assumed equal to the pressure in the boiler low-pressure circuit. The back temperature of the above steam flow in the heat box is within 50 – 60°C.

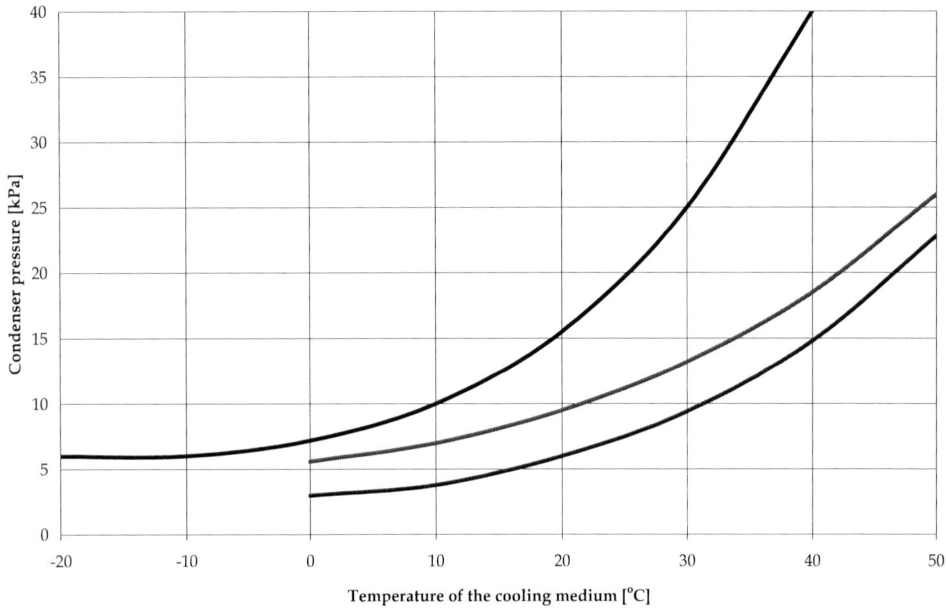

Figure 17. *Condenser pressure as a function of temperature of the cooling medium*

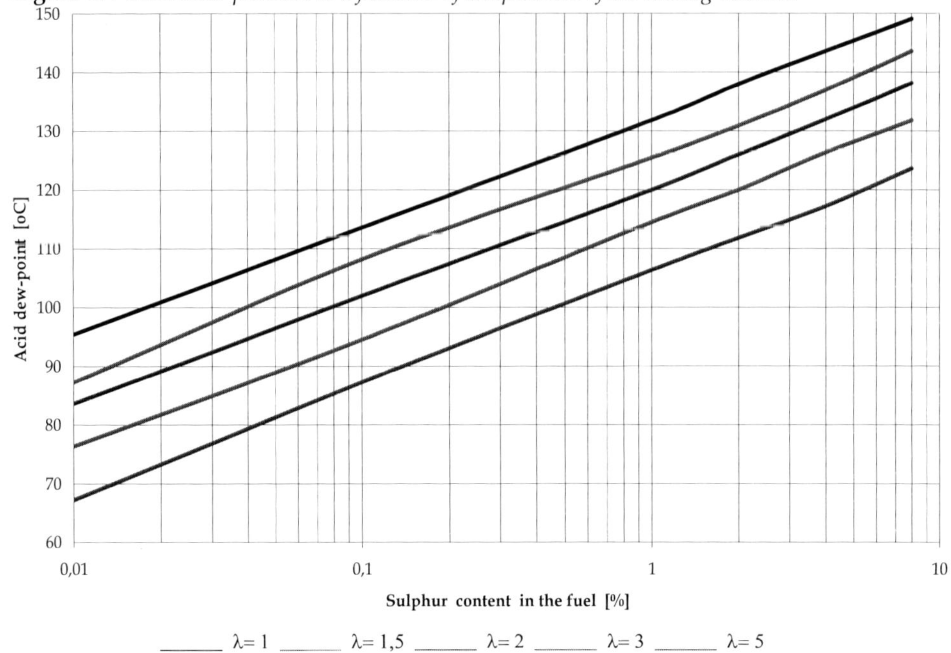

Figure 18. *Acid dew-point as a function of the sulphur content in the fuel and the excess air coefficient λ*

4.2. Optimising the steam circuit

Optimisation of the steam system is to be done in such a way so as to reach the maximal possible utilisation of the heat contained in the exhaust gas. In this sense the optimisation is reduced to selecting the steam circuit parameters for which the steam turbine reaches the highest power. The area of search for optimal steam circuit parameters is to be narrowed to the sub-area where the earlier discussed limits imposed on the steam system are met. The use of the steam system with the waste heat boiler increases the power of the propulsion system within the entire range of the main engine load.

Adding a steam turbine to the Diesel engine system increases the power of the propulsion system by $\Delta N_{ST}/N_D = 6,5 - 7,5\%$ for main engine loads ranging from 90 to 100%. The power of the steam turbine for both examined variants of power turbine supply are comparable, and slightly higher power, by about $2 - 4\%$, is obtained by the steam turbine in the variant with series power turbine supply.

The analysis of the system with an additional exchanger utilising the low-temperature waste heat from the Diesel engine to heat the condensate from the condenser before the deaerator, Fig.16, shows that the steam turbine power increases by 7- 11% with respect to that of the steam turbine without this exchanger.

The requirements concerning the waste heat boiler refer to low loss of the exhaust gas flow (which reduces the final expansion pressure in the power turbine) and small temperature concentrations (pitch points) in the boiler evaporators. There is a remarkable impact of the sulphur content in the fuel on the permissible exhaust gas temperature and the lower feed water temperature limit. In the steam turbine circuit, a minimal number of exchangers should be used (optimally: none). The optimal parameters of this circuit also depend on the piston engine load.

5. CONCLUSIONS

It is possible to implement a combined system consisting of a Diesel engine as the leading engine, a power gas turbine, and a steam turbine circuit utilising the heat contained in the Diesel engine exhaust gas. Such systems can reveal thermodynamic efficiencies comparable with combined gas turbine circuits connected with steam turbines.

5.1. Power range of combined systems

Depending on the adopted variant and the main engine load, the use of the combined system makes it possible to increase the power of the power plant by 7 to 15% with respect to the conventional power plant burning the same rate of fuel. Additional power is obtained by the system due to the recovery of the energy contained in the exhaust gas leaving the piston internal combustion engine. Thus the combined system decreases the specific fuel consumption by $6,4 - 12,8\%$ compared to the conventional power plant.

In the examined systems the power of the steam turbine is higher than that of the power turbine by 6-29%, depending on the system variant and the main engine load.

5.2. Efficiency of combined systems

The use of the combined system for ship propulsion increases the efficiency of the propulsion system, and decreases the specific fuel consumption. Additionally, it increases the propulsion power without additional fuel consumption.

Like the power, the efficiency of the combined system increases with respect to the conventional power plant by 7 to 15% reaching the level of 53 – 56% for maximal power ranges. These efficiency levels are comparable with the combined systems based on the steam/gas turbines, Fig. 1. For partial loads the efficiency curves of the combined system with the Diesel engine are more flat than those for the combined turbine systems (smaller efficiency decrease following the load decrease) .

In the combined system the maximal efficiency is reached using particular system components:

- the piston internal combustion engine with the maximal efficiency;
- Turbocharger. The turbocharger with the maximal efficiency should be used as it provides opportunities for decreasing the exhaust gas enthalpy drop in the turbine in case of the series supply variant, or exhaust gas mass flow rate in case of the parallel supply variant, which in both cases results in higher power of the power turbine;
- Power turbine. High efficiency is required to increase its power;
- Steam turbine circuit. The requirement is to obtain the maximal power of the steam turbine from the heat delivered in the exhaust gas flowing through the boiler.

5.3. Ecology

Along with the thermodynamic profits, having the form of efficiency increase, and the economic gains, reducing the fuel consumption for the same power output of the propulsion system, the use of the combined system brings also ecological profits. A typical newgeneration low-speed piston engine fed with heavy fuel oil with the sulphur content of 3% emits 17 g/kWhNOx, 12 g/kWhSOx and 600 g/kWhCO_2 to the atmosphere. The use of the combined system reduces the emission of the noxious substances by, respectively, g/kWh NOx, g/kWhSOx and g/kWhCO_2. The emission decreases by % with respect to the standard engine, solely because of the increased system efficiency, without any additional installations.

Depending on the adopted solution, the combined power plant provides oppo rtunities for reaching the assumed power of the propulsion system at a lower load of the main Diesel engine, at the same time also reducing the fuel consumption.

The article presents the thermodynamic analysis of the combined system consisting of the Diesel engine, the power gas turbine, and the steam turbine, without additional technical and economic analysis which will fully justify the application of this type of propulsion systems in power conversion systems.

6. NOMENCLATURE

b_e - specific fuel oil consumption
c_g, c_a - specific heat of exhaust gas and air, respectively
i - specific enthalpy
m – mass flow rate
N - power
p – pressure
T, t – temperature
Wu - calorific value of fuel oil
η - efficiency
$\kappa_{g,}, \kappa_a$ - isentropic exponent of exhaust gas and air, respectively

7. INDICES:

bar - air
B - barometric conditions
C - Boiler
combi - Compressor
D - combined system
d - Diesel engine
exh - supercharging
f - exhaust passage
FW - fuel
g - feet water
inlet - exhaust gas
k - inlet passage
o - parameters in a condenser
PT - live steam, calculation point
ST - Power turbine
ss - Steam turbine
T - ship living purposes
TC - Turbine
π - Turbocharger

REFERENCES

1. Dzida, M. (2009). On the possible increasing of efficiency of ship power plant with the system combined of marine diesel engine, gas turbine and steam turbine at the main engine - steam turbine mode of cooperation. Polish Maritime Research, Vol. 16, No.1(59), (2009), pp. 47-52, ISSN 1233-2585
2. Dzida, M. & Mucharski, J. (2009). On the possible increasing of efficiency of ship power plant with the system combined of marine diesel engine, gas turbine and steam turbine in case of main engine cooperation with the gas turbine fed in parallel and the steam turbine. Polish Maritime Research, Vol 16, No 2(60), pp. 40-44, ISSN 12332585

3. Dzida, M.; Girtler, J.; Dzida, S. (2009). On the possible increasing of efficiency of ship power plant with the system combined of marine diesel engine, gas turbine and steam turbine in case of main engine cooperation with the gas turbine fed in series and the steam turbine. Polish Maritime Research, Vol 16, No 3(61), pp. 26-31, ISSN 12332585
4. Kehlhofer, R. (1991). Combined-Cycle Gas & Steam Turbine Power Plants, The Fairmont Press, INC., ISBN 0-88173-076-9, USA
5. MAN B&M (October 1985). The MC Engine. Exhaust Gas Date. Waste Heat Recovery System. Total Economy, MAN BEW Publication S.A., Danish
6. MAN Diesel & Turbo (2010). Stationary Engine. Programme 4th edition, Branch of MAN Diesel & Turbo SE, Germany, Available from www.mandieselturbo.com
7. Schrott, K. H. (1995). The New Generation of MAN B&W Turbochargers. MAN BEW Publication S.A., No.236 5581E

A Study on Auxiliary Steam Reducing Device in Marine Nuclear Power Plants

Huijing Zhao, Guanhui Zhao and Rui Ma

China Ship Development and Design Center, Wuhan, Hubei Province, 430064, China

ABSTRACT

The pressure reducing device is one of the important parts of the auxiliary steam system in marine nuclear power plants. The pressure of auxiliary steam is reduced to meet the requirements of users with low pressure steam of second-loop system. This paper proposes a new auxiliary steam pressure reducing device, which can reduce the pressure of auxiliary steam in a concentrated region, and then provide low pressure steam to users. The new auxiliary steam pressure reducing device decreases the size, simplifies the adjusting mode, and improves the stability and reliability of auxiliary steam system. At last, the unsteady-state simulation based on Jtop is carried out on the auxiliary steam pressure reducing system. The results show that, when the pressure of high-pressure steam at upstream is changed, the steam pressure is stable, which is provided to users with low pressure steam.

1. INTRODUCTION

In marine nuclear power plants, the auxiliary steam system is used to transport the high-pressure steam in the header pipe to the turbogenerator, turbo-auxiliary machinery, low-pressure steam generator[1-2]. At the same time, the pressure reducing valves in the auxiliary steam system reduce the pressure of auxiliary steam, and transport the low-pressure steam to the users with low pressure steam, such as the main air ejector, auxiliary air ejector, gland air ejector, and so on. The reducing valve is one of the important parts of the auxiliary steam system, it is used to reduce the pressure of auxiliary steam to meet the requirements of users with low pressure steam[3]. Therefore, the study on pressure reducing valves in the auxiliary steam system is of great significance to the stable operating of the users with low pressure steam.

In the traditional auxiliary steam system, the pressure reducing valve and the safety valve are set on every branch connecting the steam header pipe and the main air ejector, auxiliary air ejector or gland air ejector, and so on. Although the design scheme can meet the requirements of users with low pressure steam, the auxiliary steam system has a complex piping layout, the adjusting mode is perplexity, and the size of the system is larger. As there is only one pressure reducing valve and one safety valve on every branch, the reliability of the system is low without the spare parts[4]. In the traditional auxiliary steam system, the pressure of high-pressure steam at the upstream of the pressure reducing valve is stable. However, in marine nuclear power plants, the pressure of high-pressure steam at the upstream is unstable, because of the Anti-slip characteristics. So a higher adjusting capability of pressure reducing valves is required in marine nuclear power plants.

This paper proposes a new auxiliary steam pressure reducing device in the marine nuclear power plant. The pressure reducing device reduces the pressure of high-pressure steam in a concentrated region, and then provide low pressure steam to users, for example the main air ejector, auxiliary air ejector or gland air ejector, and so on. On one hand, the design decreases the size, simplifies the adjusting mode, and improves the stability of auxiliary steam system. On the other hand, as there are two branches in the pressure reducing device: one is for use, the other is for stock, the reliability of auxiliary steam system is improved. At last, the unsteady-state simulation based on JTop is carried out on the auxiliary steam pressure reducing system to verify the characteristics of the new auxiliary steam pressure reducing device.

2. CONCEPT DESIGN

As shown in figure 1, in the traditional auxiliary steam pressure reducing device, a pressure reducing valve and a safety valve are set on every branch connecting the steam header pipe and the users with low pressure steam, including the main air ejector, auxiliary air ejector or gland air ejector, and so on. As shown in figure 2, in the new auxiliary steam pressure reducing device, the users with low pressure steam share the pressure reducing device, and the advantages of the new auxiliary steam pressure reducing device are as follows:

(1) The new auxiliary steam pressure reducing device needs less equipments, decreases the size of the system, improves the system integration, simplifies the adjusting mode, and increases the stability of the system.

(2) The new auxiliary steam pressure reducing device includes two reducing branches, and one is for use, the other is for stock. So the reliability of auxiliary steam system is improved.

(3) The adjusting capability of pressure reducing device is better. The steam pressure at the inlets of users with low pressure steam is kept stable under the condition that the pressure of high pressure steam is changed because of the Anti-slip characteristics of the marine nuclear power plant.

3. SIMULATION

3.1. Model

To study the unsteady characteristics of the new auxiliary steam pressure reducing device, this paper uses the JTopmeret software to build a graphical simulation model of the auxiliary steam pressure reducing system. Simulation model

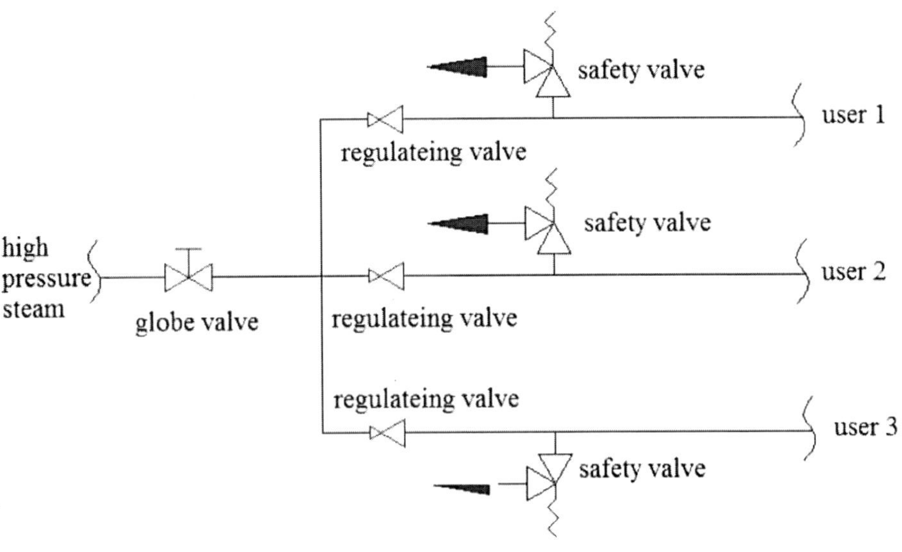

Figure 1. *Traditional auxiliary steam reducing device.*

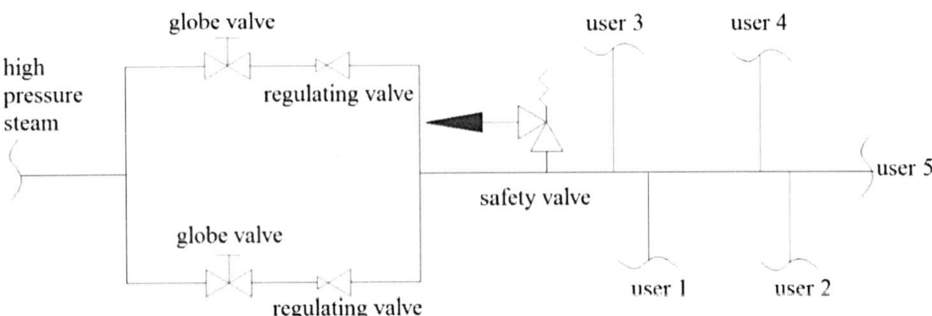

Figure 2. *New auxiliary steam reducing device.*

based on JTopmeret software is of highly consistent with the real system structure. As shown in figure 3, the auxiliary steam pressure reducing system consists of a auxiliary steam pressure reducing device and six users with low pressure steam, including two main air ejectors, an auxiliary air ejector, a gland air ejector, a sea water desalting plant and an exhaust steam system.

In order to facilitate simulate and analysis, some simplifications and presumes are made as follows:

(1) There are two branches in the real pressure reducing device: one is for use, the other is for stock. In the model, only one branch is considered.

(2) The globe valves and safety valve is not considered, which will not influence the study ressults.

(3) The resistance characteristic curve of the pressure regulating valve is assum ed to be linear[5].

(4) The heat transfer on the pipe wall is neglected.

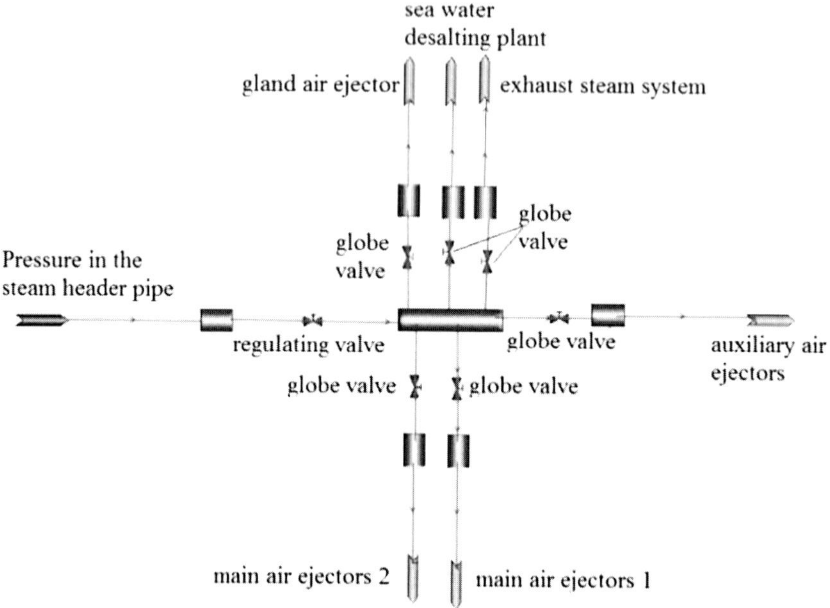

Figure 3. *The model of auxiliary steam pressure reducing system.*

Table 1. *The boundary conditions.*

Item	Number
Pressure in the steam header pipe(high pressure steam(MPa(a))	4.2
Pressure at inlet of the main air ejectors 1(MPa(a))	2.8
Mass flow rate required by the main air ejectors 1(Kg/s)	0.24
Pressure at inlet of the main air ejectors 2(MPa(a))	2.8
Mass flow rate required by the main air ejectors 2(Kg/s)	0.24
Pressure at inlet of the auxiliary air ejectors (MPa(a))	2.8
Mass flow rate required by the auxiliary air ejectors (Kg/s)	0.24
Pressure at inlet of the gland air ejector (MPa(a))	2.8
Mass flow rate required by the gland air ejector (Kg/s)	0.1
Pressure at inlet of the sea water desalting plant (MPa(a))	2.8
Mass flow rate required by the sea water desalting plant (Kg/s)	0.5
Pressure at inlet of the exhaust steam system (MPa(a))	2.8
Mass flow rate required by the exhaust steam system(Kg/s)	2.1

The simplifications and presumes above are all reasonable.

3.2. boundary conditions

As shown in table 1, at the inlet of the auxiliary steam pressure reducing system, the upstream pressure of high pressure steam is assumed. For every user, the downstream boundary condition is the mass flowrate.

3.3. The pressure control system

A control scheme of PID control is presented in this paper[6-7]. The PID control scheme logic is shown in figure 4. The control target is set as the required pressure at the inlets of users with low pressure steam. Real-time pressure of the branch is

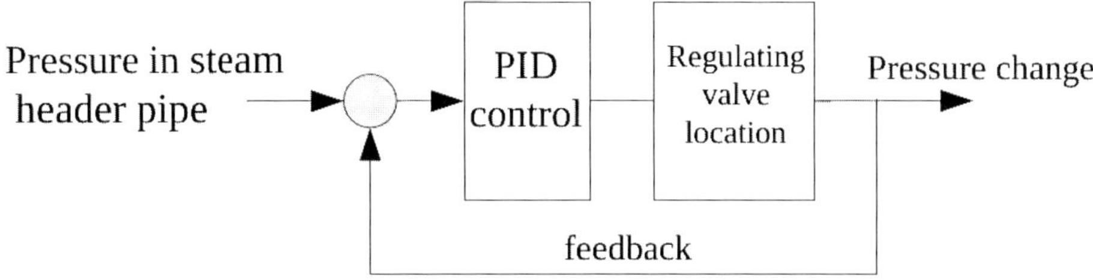

Figure 4. *PID control scheme logic.*

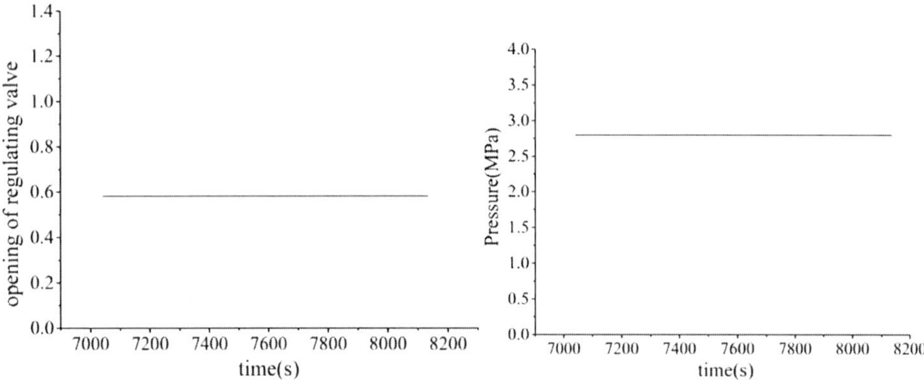

Figure 5. *The simulation result (the steam header pipe is* 4.2MPa*).*

monitored and regulating valve opening is adjusted by PID operation. When the pressure in the steam header pipe is changed, the pressure at the inlets of users with low pressure steam will change, becoming higher or lower than the required pressure. Then the signal is feedback to the PID control, and the PID operation adjusts the regulating valve opening.

4. RESULTS AND ANALYSIS

4.1. The pressure in steam header pipe is changed

To test the characteristics of the new auxiliary steam pressure reducing device, Different working conditions of the system are simulated as the pressure in steam header pipe is changed. As shown in Table 1, the pressure in the steam header pipe at design condition is 4.2MPa. The regulating valve is operating to adjust the pressure at the inlets of the users with low pressure, and the target pressure is 2.8MPa. The simulation result is present in figure 5. When the pressure in the steam header pipe is 4.2MPa, the pressure at the inlets of all users with low pressure is stable, and the target pressure (2.8MPa) is achieved. The opening of regulating valve is 0.58 (" 0 " represents the valve is closed, and "1" represents the valve is fully open), which is in the best opening range.

When the loading of second-loop system is decreased, the pressure in steam header pipe will rise. As shown in figure 6, if the pressure in the steam header pipe

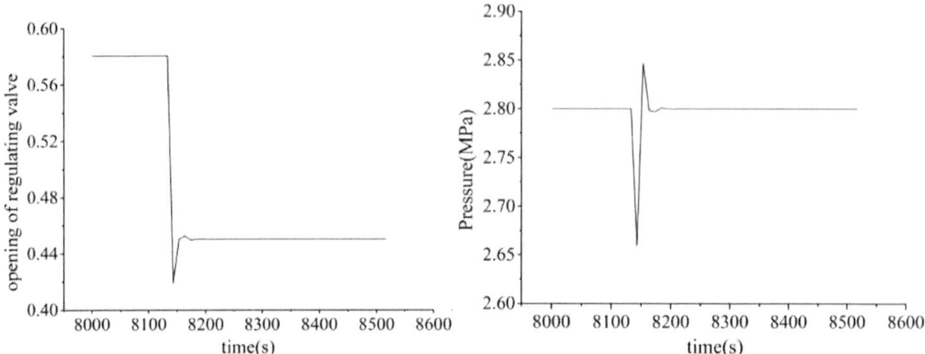

Figure 6. *The unsteady simulation result.*

rises from 4.2MPa to 4.8MPa, the opening of regulating valve and the pressure at the inlets of all users will be unstable, but reach to a stable state in a short time. The opening of regulating valve is 0.45 at the stable state(the pressure at the inlets of all users is still 2.8MPa).

If the pressure in the steam header pipe continues to rise from 4.8MPa to 5.1 MPa, the same phenomena are present as that in figure 6. The opening of regulating valve is 0.40 at the stable state. When the pressure in the steam header pipe is 5.3MPa, the opening of regulating valve is 0.38 . When the pressure in the steam header pipe is 5.5MPa, which is the highest pressure that the system can reach to, the opening of regulating valve is 0.36 . Therefore, we can draw the conclusions that the opening of regulating valve varies ranging from 0.36 to 0.58 under all the working conditions, and it is in the best opening range. When the pressure in the steam header pipe varies, the opening of regulating valve and the pressure at the inlets of all users with low pressure can reach to a stable state in a short time. The new auxiliary steam pressure reducing device is able to meet the requirements of all users with low pressure.

4.2. The mass flowrate of steam required by users with low pressure is changed

In the auxiliary steam pressure reducing system, the mass flowrate of steam required by the exhaust steam system usually varies with the loading of second-loop system. As shown in figure 7 , when the pressure in the steam header pipe is 5.3MPa, if the mass flowrate of steam required by the exhaust steam system decreases from 2.1 kg/s to 1.1 kg/s, the opening of regulating valve and the pressure at the inlets of all users can reach to a stable state in a short time. The pressure at the inlets of the two main air ejectors, an auxiliary air ejector, a gland air ejector and a sea water desalting plant is still 2.8MPa, but the pressure at the inlets of the exhaust steam system is 2.85MPa, the opening of regulating valve is 0.277. The reason is the resistance in the pipe connecting the exhaust steam system and the steam header pipe declines as the mass flowrate of steam decreased. The opening of regulating valve is 0.277 . When the mass flowrate of steam required by the exhaust steam system is 0.2 kg/s, the opening of regulating valve is 0.173, which is out of the best opening range.

When the pressure in the steam header pipe is 5.2MPa, and the mass flowrate of steam required by the exhaust steam system is 0.2 kg/s, the opening of regulating valve is 0.34 . If the mass flowrate of steam required by the exhaust steam system is 0 kg/s, the opening of regulating valve is 0.3 . The results show that the regulating valve is able to meet the requirements of all users, except for the working condition that the pressure in the steam header pipe is 5.5MPa.

5. CONCLUSIONS

The new auxiliary steam pressure reducing device decreases the size, simplifies the adjusting mode, and improves the stability and reliability of auxiliary steam system.

When the pressure in steam header pipe is changed, the new auxiliary steam pressure reducing device is able to meet the requirements of all users with low pressure.

When the mass flowrate of steam required by users with low pressure is chang ed, the pressure at the inlets of users with low pressure steam can't be the target value(2.8MPa). At the working condition that the pressure in the steam header pipe is 5.5MPa, the regulating valve isn't able to meet the requirements of all users.

REFERENCES

1. Hui, F. J. (2017) Simulation Study on Pipe Network Characteristics of Second Loop Main Auxiliary Steam System. Harbin: Harbin Engineering University. (in Chinese)
2. Liu, X. C., Cai, R. Z., Lu, C. D. (2002) Simulation investigation of large-scale steam heating network. Journal of system simulation, 14(3):397-399.
3. Tang, L. R. (2011) Research on Marine Auxiliary Steam System. Naval Architecture and Ocean Engineering, 2: 42-46.
4. Ye, X. Z. (2013) Study on Adjustment Characteristic and Maintenance Strategy of Reducing Valve in Auxiliary Steam System. Harbin: Harbin Engineering University. (in Chinese)
5. Shen,Y. F. (2011) The selection and simulation research for the control valve of steam system of a smeltery. Wuhan:China Ship Development and Design Center. (in Chinese)
6. Guimaraes, L. N. F., Oliverira, N. D., Borges, E. M., et al. (2008) Derivation of a nine variable model of a U-tubesteam generator coupled with a three-element controller. Applied Mathematical Modeling, 32(6): 1027-1043.
7. Saury, D., Harmand, S., Siroux, M. (2005) Flash evaporation from a water pool: influence of the liquid height and of the depressurization rate [J].International Journal of Thermal Sciences, 44(10): 953-965.

CHAPTER

9

Power Tracking Control of Marine Boiler-Turbine System Based on Fractional Order Model Predictive Control Algorithm

Shiquan Zhao[1], Sizhe Wang[1], Ricardo Cajo[2], Weijie Ren[1,3] and Bing Li[1]

[1]*College of Intelligent Systems Science and Engineering, Harbin Engineering University, Harbin 150001, China*
[2]*Facultad de Ingeniería en Electricidad y Computación, Escuela Superior Politécnica del Litoral, ESPOL, Campus Gustavo Galindo Km 30.5 Vía Perimetral, P.O. Box 09-01-5863, Guayaquil 090150, Ecuador*
[3]*Faculty of Electronic Information and Electrical Engineering, Dalian University of Technology, Dalian 116024, China*

ABSTRACT

The marine boiler-turbine system is the core part for the steam-powered ships with complicated dynamics. To improve the power tracking performance and fulfill the requirement of high utilization rate of fossil energy, the control performance of the system should be improved. In this paper, a nonlinear model predictive control method is proposed for the boiler-turbine system with fractional order cost functions. Firstly, a nonlinear model of the boiler-turbine system is introduced. Secondly, a nonlinear extended predictive self adaptive control(EPSAC) method is designed to the system. Then, integer order cost function is replaced with a fractional order cost function to improve the control performance, and also the configuration of the cost function is simplified. Finally, the superiority of the proposed method is proved accordring to the comparison experiments between the fractional order model predictive control and the traditional model predictive control.

Keywords: *boiler-turbine; nonlinear model predictive control; fractional order calculus; distributed control*

1. INTRODUCTION

In order to reduce the waste of fossil energy and CO_2 emissions, many countries have released different policies. China has released a policy document to fulfill its target of reaching peak carbon emissions by 2030. The United States of America released policy to cut carbon emissions in half by 2030. A lot of renewable power technologies were also proposed. However, most applied energy still comes from the combustion of fossil fuels. In addition, the control performance has a close relationship with the utilization of fossil fuels [1]. In this paper, the fossil fuels in ships are focused. Many of the ships are equipped with internal combustion engine, or gas turbine. However, for large scale ships, the power systems based on boiler-turbine still occupy a large proportion [2-4]. For example, many aircraft carriers are powered with boiler-turbine system. Hence, a lot of academics and companies are doing research to improve the control performce of the boiler-turbine system [5-9], of which there are three manipulated variables and three controlled variables with complicated dynamics. For the marine steam power plant, the disturbance and the energy required changes more frequently compared than that on land. However, there is not so much research about the control for the marine boiler-turbine system.

In the boiler-turbine system, the interactions of rvariables and constraints are the mainly reasons which make it difficult to obtain a satisfied control performance. The input variables for the system are the flow rates of fuel, steam to the turbine, and feedwater to the boiler, while the output variables are the steam pressure in the drum of the boiler, power required of the turbine, and the water level of the drum. The constraints are the limitations for the actuator, including the upper and lower bound, and rate limiter. In addition, power requirement changes a lot, which is usually treated as disturbance. In order to compensate unknown disturbances in the boiler-turbine system, a high order sliding mode observer was designed for a baseline exponentially stable feedback controller [9]. To improve the economy of the boiler-turbine system, the economic index was utilized directly in the cost function, and a global economic optimum routine was obtained for the system [10]. In the literature [11], the model of the boiler-turbine system was linearized and decoupled with an adaptive feedback linearization method, and a second order sliding mode controller was designed to deal with the disturbances and uncertainties. Ref. [12] applied an online policy iteration integral reinforcement learning method to the boiler-turbine system, and optimal tracking control performance was obtained.

The model predictive control(MPC) has the advantages in dealing with the nonlinear dynamics, interactions and constraints problems [13], hence, MPC is a preferred choice for researchers. The MPC is studied in many fields such as building energy management [14], landscape office lighting regulatory system [15,16], wind turbines [17], tank-system [18], pressure oscillation adsorption process [19,20], permanent-magnet synchronous motor [21], autonomous underwater vehicle [22], and so on. For the boiler-turbine system, there are also some applications of MPC. A nonlinear model predictive control method was designed for boiler-turbine system with a data driven model [23], and the optimal problem was solved by immune genetic algorithm. Ref. [24] presented a zone economic model predictive cotroller to fulfill the economic target of the boiler-turbine system. Fuzzy model predictive control was designed to realize load tracking and economy of the boiler-turbine system, and a fuzzy model was used to approximate the nonlinear dynamics of the system [25]. Other MPC applications on boiler-turbine system can be found in [26-28].

For the MPC method, the weighting factors have a significant different effect on the control performance, and the number of the weighting factors is large, which makes it diffcult to obtain a good choice for these parameters. For example, if the N_{ip} denotes the prediction horizon for the i th output, and N_{jc} for the j th control horizon, the number of weighting factor will be $\sum_{i=1}^{n} N_{ip} + \sum_{j=1}^{m} N_{jc}$, where m and n are the numbers of output and input. The commonly used method to tune the weighting factors is trial and error or choose them empirically. Another alternative way to choose the control parameters is optimization method [29,30]. However, due to the high dimension of the weighting factors, it is difficult to optimize them. So, in this paper, the fractional order model predictive control(FOMPC) for the boiler-turbine system is proposed. By two fractional order papermeters (one for the tracking error, and another for the control effort), the weight factors matrices can be obtained, which reduces the difficulties in weighting factors configuration. The boilerturbine system is a nonlinear multiple inputs and multiple outputs system, so the nonlinear distributed structure of MPC is studied for the system.

The rest of the paper is structured as follows. The boiler-turbine system is formulated in Section 2; Section 3 presents the nonlinear distributed MPC with EPSAC framework. Fractional order MPC is designed for the boiler-turbine system in the Section 4. The simulation experiments are presented in Section 5. The last section gives the conclusions.

2. BOILER-TURBINE SYSTEM

The boiler-turbine system is a core part in the power plant, and Figure 1 shows the structure of the system [9]. The details elements are indicated in the figure, and they are listed as follows: 1-drum; 2-superheater; 3-water spray desuperheater; 4-valve for the steam to turbine; 5-turbine high-pressure cylinders; 6-foward control valve; 7-backward control valve; 8-turbine middle- and low-pressure cylinders; 9-shafting; 10-condenser; 11-replenish water; 12-condensate pump; 13-low-pressure heater; 14-deaerator; 15-feed water pump; 16-high-pressure heater; 17-feed water valve; 18-economizer; 19-downcomers; 20-water-cooled walls; 21-furnace; 22-heat flow control valve; 23-nozzle; 24-blower; 25-preheater for air; 26-air conditioner; 27-flue gas baffle; 28-induced draft fan; 29-flue; 30-gearbox; 31-turbine.

Due to the lack of data for marine boiler-turbine sytem, it is difficult to obtain its model. The structure is similar to that on land. Hence, a model of on load power plant is chosen for study, and the nonlinear model of the boiler-turbine system is shown as follows according to the literature [31]:

$$\dot{x}(t) = F(x(t)) + G(x(t))u(t), \tag{1}$$

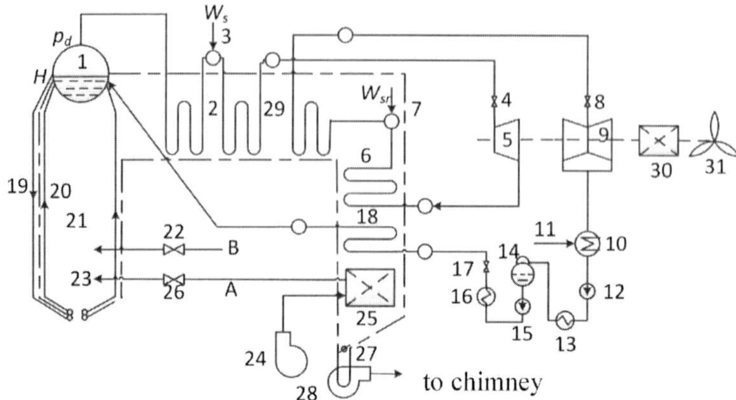

Figure 1. *The structure of a boiler-turbine unit.*

and the $F(x(t)), G(x(t))$ are defined as follows:

$$F(x(t)) = \begin{bmatrix} 0 \\ -0.1x_2 - 0.016x_1^{9/8} \\ 0.0022x_1 \end{bmatrix} \tag{2}$$

$$G(x(t)) = \begin{bmatrix} 0.9 & -0.0018x_1^{9/8} & -0.15 \\ 0 & 0.073x_1^{9/8} & 0 \\ 0 & -0.0129x_1 & 1.6588 \end{bmatrix} \tag{3}$$

where the inputs $u = [u_1, u_2, u_3]^T$ for the system are the valve opening of fuel, steam to the turbine and feedwater to the drum. The states are drum steam pressure, power required for the turbine and steam water density denoted by $x = [x_1, x_2, x_3]^T$. The outputs are drum steam pressure, power required for the turbine and water level in the drum. The level of the drum can be calculated as:

$$L = 0.05\,(0.13073x_3 + 100\alpha_s + q_e/9 - 67.975) \tag{4}$$

and $q_e = (0.854u_2 - 0.147)\,x_1 + 45.59u_1 - 2.51u_3 - 2.096$; $\alpha_s = \frac{(1-0.001538x_3)(0.8x_1-25.6)}{x_3(1.0394-0.0012304x_1)}$. The q_e and α_s denote the evaporation rate and steam quality, respectively.

The rates and amplitudes limitaion for the inputs are listed as follows:

$$\begin{cases} -0.007 \le \frac{du_1}{dt} \le 0.007 & 0 \le u_1 \le 1 \\ -2.0 \le \frac{du_2}{dt} \le 0.02 & 0 \le u_2 \le 1 \\ -0.05 \le \frac{du_3}{dt} \le 0.05 & 0 \le u_3 \le 1 \end{cases} \tag{5}$$

For the boiler-turbine, there are different operating points. To evaluate the effectiveness of the proposed method, experiments around the following operating points shown in Table 1 are carried out.

Table 1. *Operating points for the boiler-turbine system.*

Operating Point	Pressure	Power	Density
1	99.3 kg/m^2	80.9 MW	396
2	120 kg/m^2	110 MW	331

The drum water level should always be kept as zero meters.

3. NONLINEAR DISTRIBUTED MPC FOR THE BOILER-TURBINE SYSTEM

According to the introduction of the boiler-turbine, it can be found that this system is a nonlinear multiple inputs multiple outputs system. Hence, the nonlinear MPC is designed for the system with distributed structure.

3.1. The Basic of the EPSAC

This part presents the basic of EPSAC. For more details about the EPSAC, it can be found in the refs. [32-34]. For a discrete system, the system output can be expressed as:

$$y(t) = x(t) + w(t) \tag{6}$$

where $y(t)$ is the system output; $x(t)$ is the model output and the $w(t)$ is the disturbances. $x(t)$ can be calculated according to the model of the system as follow:

$$x(t) = f[x(t-1), x(t-2), \ldots, u(t-1), u(t-2), \ldots] \tag{7}$$

In Equation (7), the $f(x,u)$ denotes the model of the system, $x(t-i)$ and $u(t-i)$ $i = 1, 2, \ldots$ indicate the past model outputs and inputs.

In the EPSAC, the input scenario for the future is composed with two parts:

$$u(t+k \mid t) = u_{base}(t+k \mid t) + \delta u(t+k \mid t) \tag{8}$$

where the $u_{\text{base}}\,(t+k \mid t)$ and $\delta u(t+k \mid t)$ are the basic and optimized future control actions. Then the future system output can be predicted as:

$$y(t+k \mid t) = y_{base}(t+k \mid t) + y_{opt}(t+k \mid t) \tag{9}$$

where $y_{base}(t+k \mid t)$ is the result of the basic future control action; $u_{base}(t+k \mid t)$ and $y_{\text{opt}}\,(t+\, k \mid t)$ can be calculated according to the optimized future control action $\delta u(t+k \mid t)$.

The $y_{opt}(t+k \mid t)$ can be obtained with:

$$y_{opt}(t+k \mid t) = h_k \delta u(t \mid t) + h_{k-1} \delta u(t+1 \mid t) + \ldots + g_{k-N_c+1} \delta u\,(t+N_c-1 \mid t) \tag{10}$$

In Equation (10), the h_i and g_i are the impulse response and step response coefficients of the system, respectively; N_c and N_p are the control horizon and the prediction horizon, respectively. The system output can be re-written in matrix form:

$$\mathbf{Y} = \overline{\mathbf{Y}} + \mathbf{GU} \tag{11}$$

where $\mathbf{Y} = \left[y\left(t+N_1 \mid t\right) \ldots y\left(t+N_p \mid t\right)\right]^T, \mathbf{U} = \left[\delta u(t \mid t) \ldots \delta u\left(t+N_c-1 \mid t\right)\right]^T$ $, \overline{\mathbf{Y}} = \left[y_{\text{base}}\left(t+ \ N_1 \mid t\right) \ldots y_{\text{base}}\ \left(t+N_P \mid t\right)\right]^T$; N_1 is the time delay of the system, and

$$\mathbf{G} = \begin{bmatrix} h_{N_1} & h_{N_1-1} & \cdots & g_{N_1-N_c+1} \\ h_{N_1+1} & h_{N_1} & \cdots & \cdots \\ \cdots & \cdots & \cdots & \cdots \\ h_{N_P} & h_{N_P-1} & \cdots & g_{N_P-N_c+1} \end{bmatrix}$$

The disturbance term $w(t)$ in (6) includes all the effects on the system output. It can be modeled by a colored noise process as:

$$w(t+k \mid t) = \frac{C\left(q^{-1}\right)}{D\left(q^{-1}\right)} w_f(t+k \mid t) \tag{12}$$

where q^{-1} is the backward shift operator.

In this work, the $C\left(q^{-1}\right)$ and $D\left(q^{-1}\right)$ are designed as follows:

$$\frac{C\left(q^{-1}\right)}{D\left(q^{-1}\right)} = \frac{1}{\left(1-q^{-1}\right)\left(1-ae^{+j\alpha}q^{-1}\right)\left(1-ae^{-j\alpha}q^{-1}\right)} \tag{13}$$

with $\alpha = 2\pi f_0 T_s$ and $a \approx 1$; T_s is the sampling time and $a \leq 1$ for stability.

The cost function for the boiler-turbine system can be defined as:

$$J_{MPC} = \sum_{k=N_1}^{N_2} p_k[r(t+k \mid t) - y(t+k \mid t)]^2 + \sum_{k=1}^{N_u} q_k \Delta u(t+k)^2 \tag{14}$$

The p_k and q_k are nonnegative weighting factors, and they are usually kept as constants. The matrix form of Equation (14) can be written as:

$$J_{MPC} = (\mathbf{R}-\mathbf{Y})^T\mathbf{P}(\mathbf{R}-\mathbf{Y}) + \mathbf{U}^T\mathbf{Q}\mathbf{U} = (\mathbf{R}-\overline{\mathbf{Y}}-\mathbf{G}\mathbf{U})^T\mathbf{P}(\mathbf{R}-\overline{\mathbf{Y}}-\mathbf{G}\mathbf{U}) + \mathbf{U}^T\mathbf{Q}\mathbf{U} \tag{15}$$

with $P = \text{diag}\left(p_1, p_2, \ldots, p_{(N_2-N_1+1)}\right)$ and $Q = \text{diag}\left(q_1, q_2, \ldots, q_{N_u}\right)$

For systems with constraint, the optimization problem can be solved with qua dratic programming. Otherwise, the results of the optimal input part, which are indicated by $\delta u(t+k \mid t)$, can be obtained as:

$$\mathbf{U}^*_{\text{MPC}} = \left(\mathbf{G}^T\mathbf{P}\mathbf{G} + \mathbf{Q}\right)^{-1} \mathbf{G}^T\mathbf{P}(\mathbf{R}-\overline{\mathbf{Y}}) \tag{16}$$

3.2. The Fractional Order MPC

For the fractional order MPC, the cost function is designed as:

$$J_{FOMPC} = {}^{\gamma}I_{N_1}^{N_2} p_k[r(t+k \mid t) - y(t+k \mid t)]^2 + {}^{\lambda}I_1^{N_c} q_k \Delta u(t+k)^2 \tag{17}$$

where ${}^{\gamma}I_{N_1}^{N_2}$ and ${}^{\lambda}I_1^{N_c}$ indicate fractional order integral with fraction order of γ and λ; $[N_1, N_2]$ and $[1, N_c]$ are the integration intervals.

According to [35], the Equation (17) can be written by:

$$\begin{aligned} J_{FOMPC} &= (\mathbf{R}-\mathbf{Y})^T\mathbf{P}\boldsymbol{\Gamma}(T_s,\gamma)(\mathbf{R}-\mathbf{Y})+\mathbf{U}^T\mathbf{Q}\boldsymbol{\Lambda}(T_s,\lambda)\mathbf{U} \\ &= (\mathbf{R}-\overline{\mathbf{Y}}-\mathbf{GU})^T\mathbf{P}\boldsymbol{\Gamma}(T_s,\gamma)(\mathbf{R}-\overline{\mathbf{Y}}-\mathbf{GU})+\mathbf{U}^T\mathbf{Q}\boldsymbol{\Lambda}(T_s,\lambda)\mathbf{U} \end{aligned} \tag{18}$$

$$\boldsymbol{\Gamma}(T_s,\gamma) = T_s^{\gamma}\,\mathrm{diag}\left(m_{N_2-N_1}, m_{N_2-N_1-1},\ldots,m_1,m_0\right) \tag{19}$$

$$\boldsymbol{\Lambda}(T_s,\lambda) = T_s^{\lambda}\,\mathrm{diag}\left(m_{N_c}, m_{N_c-1},\ldots,m_1,m_0\right) \tag{20}$$

The m_i in Equations (19) and (20) with fractional order α can be calculated as:

$$m_j = \omega_j^{(-\alpha)} - \omega_{j-n}^{(-\alpha)} \tag{21}$$

where n is the number of the m_i, and ω can be calculated with:

$$\omega_j^{-\alpha} = \begin{cases} (1-(1-\alpha)/j)\omega_{j-1}^{(-\alpha)} & j>0 \\ 1 & j=0 \\ 0 & j<0 \end{cases} \tag{22}$$

According to Equation (19) to Equation (22), the weight matrix in the cost function J_{FOMPC} can be easily tuned with fractional order γ and λ. For system with constraints, the optimization for input sequence can be solved with quadratic programming. For the case without constraints, the results for the optimal input $\delta u(t+k \mid t)$ can be calculated as:

$$\mathbf{U}_{\mathbf{FOMPC}}^{*} = \left(\mathbf{G}^T\mathbf{P}\left(\boldsymbol{\Gamma}+\boldsymbol{\Gamma}^{\mathrm{T}}\right)\mathbf{G}+\mathbf{Q}\left(\boldsymbol{\Lambda}+\boldsymbol{\Lambda}^{\mathrm{T}}\right)\right)^{-1}\mathbf{G}^T\mathbf{P}\left(\boldsymbol{\Gamma}+\boldsymbol{\Gamma}^{\mathrm{T}}\right)(\mathbf{R}-\overline{\mathbf{Y}}) \tag{23}$$

3.3. Application of the Fractional Order EPSAC to the Nonlinear MIMO System with Distributed Structure

The fractional order MPC introduced above is for the linear case. In order to apply the FOMPC to the boiler-turbine system, the nonlinear FOMPC is studied. According to the Equation (8) and (9), the principle of superposition is applied for linear system. To get over the superposition, the optimal future input $\delta u(t+k \mid t)$ should be removed iteratively smaller tends to zero [36]. The procedure for the nonlinear MPC is summarized as follows:

- Choose an initial base input sequence $u_{\text{base}}(t+k \mid t), k = 0\ldots N_u - 1$, this part should be as close as possible to the optimal input $u(t+k \mid t)$ to make the $\delta u(t+k \mid t)$ close to zero, which means that the term $y_{\text{opt}}(t+k \mid t)$ equals to zero;
- After chooseing the base future input, the $\delta u(t+k \mid t)$ can be calculated. The $\delta u(t+k \mid t)$ is not close to zero at the moment;

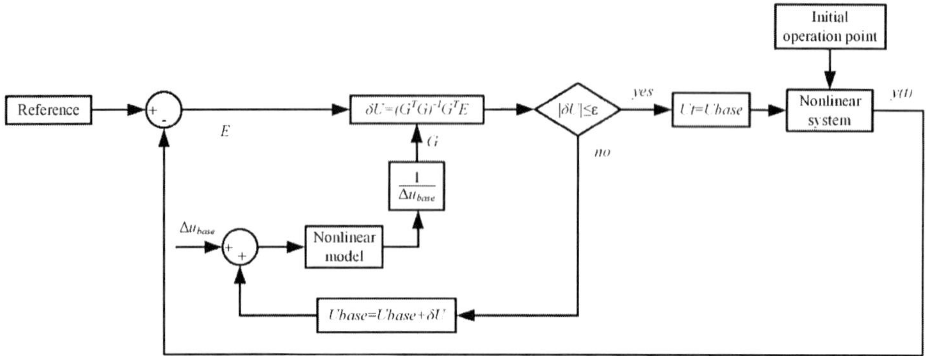

Figure 2. *The flow chart of nonlinear MPC.*

Table 2. *Parameters for the MPC*

Parameters	N_c	T_s	N_p	N_1	N_s
Values	$N_{c1} = 1$, $N_{c2} = 1$, $N_{c3} = 1$ samples	5s	$N_{p1} = 15$, $N_{p2} = 15$, $N_{p3} = 15$ samples	1	100

- Take the $u(t + k \mid t)$ from the second step as the new $u_{\text{base}}\ (t + k \mid t)$, and calculate $\delta u(t{+}\ k \mid t)$ again.
- Repeat step 2 and 3 until the $\delta u(t + k \mid t)$ is as close as possible to zero, then the $u_{\text{base}}\ (t + k \mid t)$ can be applied to the system at the time $t + 1$.

The flow chart of nonlinear MPC is shown in Figure 2.

The boiler-turbine is a MIMO system, and there are strong interactions between variables. In order to calculate the optimal input sequence, the effect from coupling variables should be considered and the communication network should be established. In this work, the distributed structure is applied. The pseudocode is provided in Algorithm 1.

The boiler-turbine system is a nonlinear MIMO system, hence, the nonlinear MPC with distributed scheme will be applied. According to the procedure of nonlinear MPC and algorithm for the distributed MPC, the optimal future input sequence should fulfill the follow conditions:

$$\begin{cases} \left\| \mid \delta \mathbf{U}_i^{i\ \text{ter}\ +1} - \delta \mathbf{U}_i^{i\ \text{ter}} \right\| \leqslant \varepsilon_i) \\ \delta \mathbf{U}_i^{i\ \text{ter}\ +1} \approx 0 \end{cases} \tag{24}$$

4. SIMULATION OF THE FRACTIONAL ORDER MPC ON BOILER-TURBINE SYSTEM

This section shows the simulation results of the fractional order MPC. Firstly, different fractional order terms are applied to different loops. Then, the best fractional order terms are applied to the drum steam pressure loop, power for turbine

Algorithm 1 Algorithm for the distributed MPC

1: The loop i receives an optimal local control action $\delta\mathbf{U}_i$ for the first time, which will be marked as $iter = 0$, and the local control action $\delta\mathbf{U}_i$ can be marked as $\delta\mathbf{U}_i^{iter}$, where $\delta\mathbf{U}_i$ indicates the vector of the optimizing future control actions with length of N_{ci};

2: The information of coupling variables $\delta\mathbf{U}_j^{iter}$ $(j \in N_i, N_i = \{j \in N : \mathbf{G}_{ij} \neq 0\})$ will be sent to the loop i, and the $\delta\mathbf{U}_i^{iter+1}$ will be recalculated with the information of $\delta\mathbf{U}_j^{iter}$ from other loops;

3: The termination condition can be designed as: $(||\delta\mathbf{U}_i^{iter+1} - \delta\mathbf{U}_i^{iter}|| \leqslant \varepsilon_i) \vee (iter + 1 > \overline{iter})$. where ε is a positive constant and $\overline{iter}$ indicates the upper bound of the number of iteration times. If the termination condition is reached, the $\delta\mathbf{U}_i^{iter+1}$ will be adopted to the system. Otherwise, $iter = iter + 1$, and return to the Step 2.

4: The final optimal control effort can be obtained as $\mathbf{U}_t = \mathbf{U}_{base} + \delta\mathbf{U}^{iter}$, which will be applied to the system;

5: $t = t + 1$, return to Step 1.

Table 3. *Fractional order terms for each loop.*

Loops	Fractional Order Terms
Drum steam pressure loop	[0.5, 1, 1.5, 3]
Power loop	[0.8, 1, 1.2, 1.5]
Drum water level loop	[0.5, 1, 1.7, 2, 2.5]

loop and water level loop, and the results are compared with the integer order MPC. Finally, the results are discussed.

The parameters configuration are listed in Table 2.

In Table 2, N_{ci} and $N_{pi}(i = 1,2,3)$ are control horizon and prediction horizon, respectively; N_s is the number of simulation steps. The termination conditions for the nonlinear iteration are set as: $\delta\mathbf{U}_i^{iter} \leqslant 0.05$; or the iteration times iter $_{\text{nmpc}}$ > 5. The termination conditions for the distributed MPC are set as: $\left\|\delta\mathbf{U}_i^{i\text{ ter}+1} - \delta\mathbf{U}_i^{i\text{ ter}}\right\| \leqslant 0.005$; or iter > 5.

4.1. The Influence of Fractional Order Terms to the Different Loops

In order to test the effect of different fractional order on the control performance, different fractional order terms are introduced to the cost function for each loop. The details for fractional order terms are listed in Table 3. In this work, the γ for the reference tracking and λ for the control effort are chosen the same for simplification.

According to the results shown in Figures 3-5, it can be seen that the effectiveness of different fractional order terms varies a lot. For the drum steam pressure control, the best fractional order term is 1.5 ; for the required power control, it is 1; and for the drum water level control, the fractional order term of 1.7 is the best.

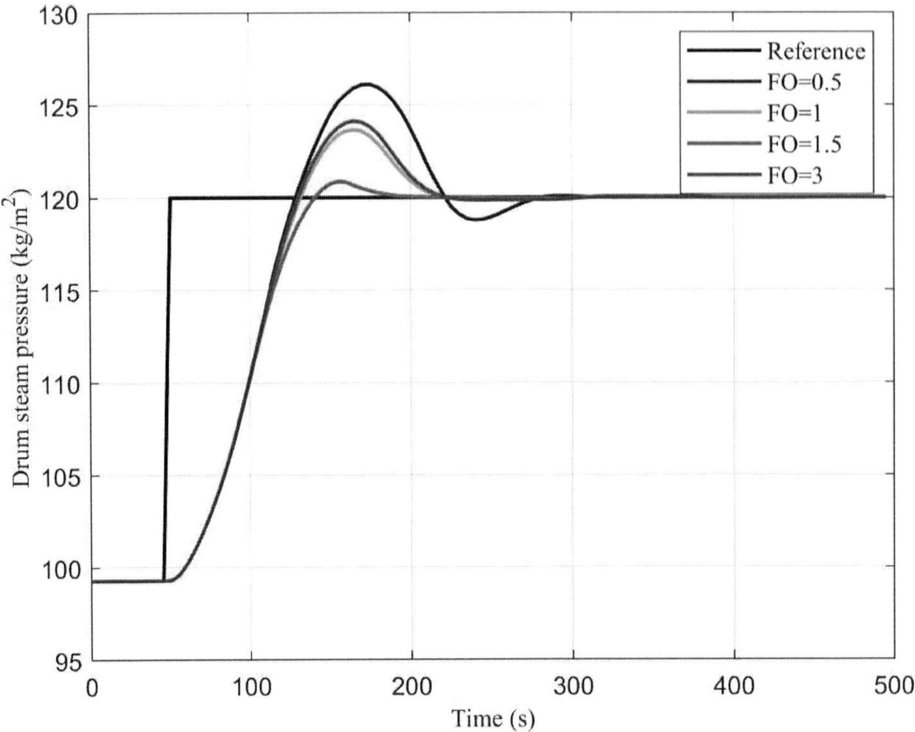

Figure 3. *The drum steam pressure with different fractional orders.*

4.2. The Influence of Fractional Order Terms to the Different Loops

This section shows the comparison experiment between FOMPC and traditional MPC. The fractional order terms applied in the boiler-turbine system are chosen as 1.5, 1,1.7 for the three loops, respectively.

In order to evaluate the effectiveness of the proposed method, the following performance indexes are compared, including Integrated Absolute Relative Error (IARE), Integral Secondary control output (ISU), Ratio of Integrated Absolute Relative Error (RIARE), Ratio of Integral Secondary control output (RISU) and combined index (J).

$$\text{IARE}\, E_i = \sum_{k=0}^{N_s-1} |r_i(k) - y_i(k)| \,/ r_i(k) \quad (i = 1,2,3) \tag{25}$$

$$ISU_i = \sum_{k=0}^{N_s-1} (u_i(k) - u_{ssi}(k))^2 \quad (i = 1,2,3) \tag{26}$$

$$\text{RIARE}_i\,(C_2, C_1) = \frac{\text{IARE}_i\,(C_2)}{\text{IARE}_i\,(C_1)} \quad (i = 1,2,3) \tag{27}$$

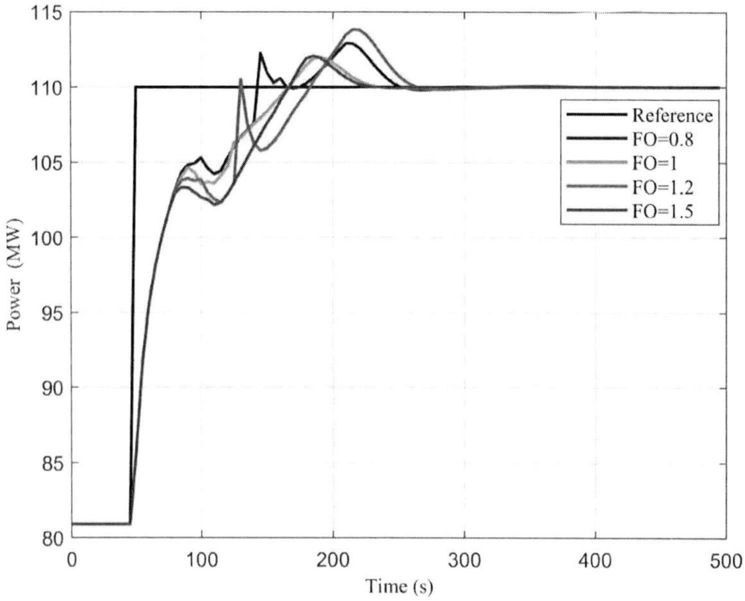

Figure 4. *The required power for turbine with different fractional orders.*

Table 4. *Performance indexes for IARE and ISU in the power reference tracking experiment.*

Index		Drum Steam Pressure	Power	Drum Water Level
IARE	MPC	2.0072	1.5837	3.5389
	FOMPC	1.8342	1.4955	3.4534
ISU	MPC	1.1961	0.1068	18.8118
	FOMPC	0.9444	0.1042	17.1870

$$\mathrm{RISU}_i\,(C_2, C_1) = \frac{\mathrm{ISU}_i\,(C_2)}{\mathrm{ISU}_i\,(C_1)} \quad (i = 1, 2, 3) \tag{28}$$

$$J\,(C_2, C_1) = \frac{1}{3}\sum_{i=1}^{3} \frac{w_1\,\mathrm{RIARE}_i\,(C_2, C_1) + w_2\,\mathrm{RISU}_i\,(C_2, C_1)}{w_1 + w_2} \tag{29}$$

where u_{ssi} is the steady state value of i th input; C_1, C_2 are the two compared controllers; the weighting factors w_1 and w_2 in equation (29) are chosen as $w_1 = w_2 = 0.5$.

The combined index J for the reference tracking case is 0.6564 . From Figure 6, Tables 4 and 5, it shows that the IARE and ISU of fractional order MPC are both smaller than that of traditional MPC, which means the fractional order MPC can obtain better performance with less control effort changes. By choosing suitable fractional order terms, the fractional order MPC shows superiority compared with integer order MPC.

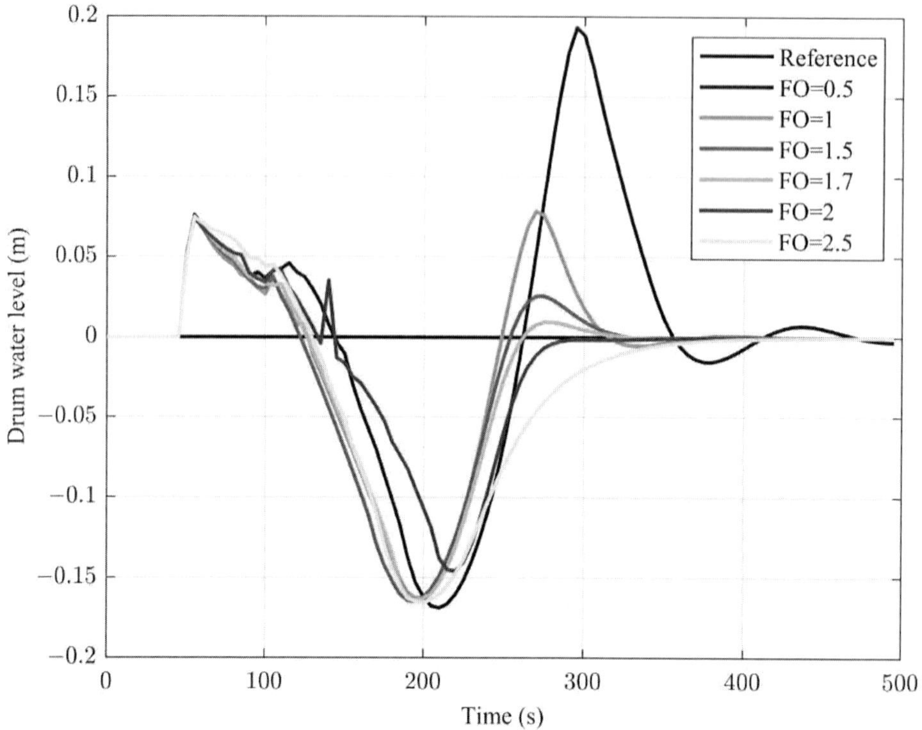

Figure 5. *The drum water level with different fractional orders.*

Table 5. *Performance indexes for RIARE and RISU in the power reference tracking experiment (MPC is the C_2 and FOMPC the C_1 according to Equations (27) and (28)).*

Index		Drum Steam Pressure	Power	Drum Water Level
IARE	MPC	2.0072	1.5837	3.5389
	FOMPC	1.8342	1.4955	3.4534
ISU	MPC	1.1961	0.1068	18.8118
	FOMPC	0.9444	0.1042	17.1870

5. CONCLUSIONS

This paper proposed a fractional order model predictive controller for the boilerturbine system. Due to the nonlinearity and multiple variables of the boiler-turbine, the nonlinear MPC with distributed scheme is designed, and the termination conditions are given. The integer order cost function is replaced with the fractional order cost function, which simplified the configration of the weighting factor matrices in the cost function. The number of weighting factors required to be tuned decreases from $N_p + N_c$ to two. According to the simulation for power tracking, it is proved that the fractional order MPC improves the control performance compared with the traditional MPC method. In this work, better control performance is obtained with fractional order MPC; however, how the fractional

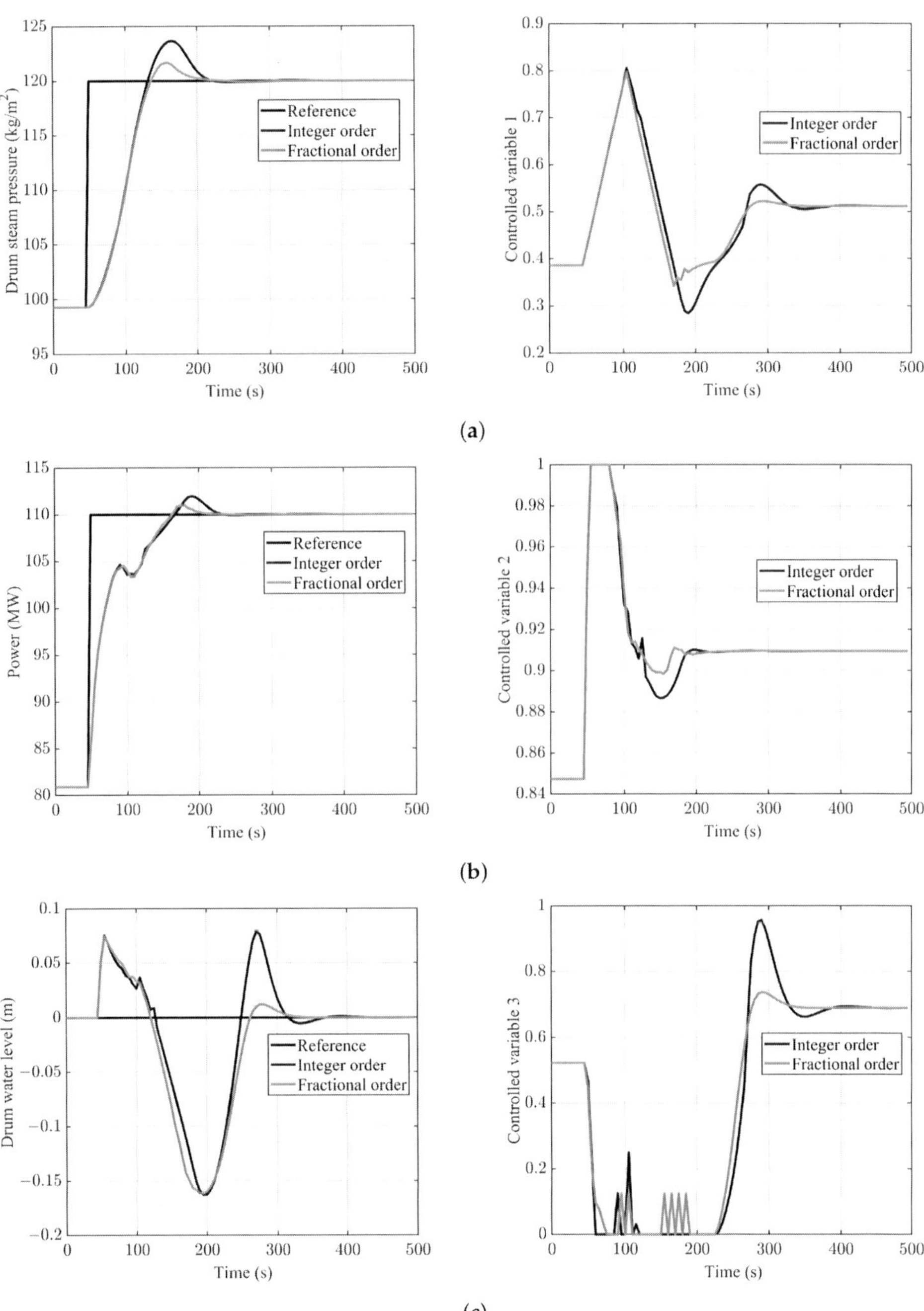

Figure 6. *Outputs and inputs of the boiler-turbine system with FOMPC and MPC in the reference tracking experiment (the outputs are listed on the left hand, and the inputs are listed on the right hand). (a) control loop for drum steam pressure, (b) control loop for required power, (c) control loop for drum water level.*

order effects the control performance is not clear, which can be researched further in the future.

Author Contributions

Methodology, S.Z.; software, S.W.; validation, S.W., R.C. and W.R.; formal analysis, S.W., R.C. and B.L.; writing-original draft preparation, S.Z.; writing-review and editing, S.Z., S.W., R.C., W.R. and B.L.; supervision, S.Z.; project administration, S.Z. and B.L.; funding acquisition, S.Z. and B.L. All authors have read and agreed to the published version of the manuscript.

REFERENCES

1. Yang, F.; Tang, C.; Antonietti, M. Natural and artificial humic substances to manage minerals, ions, water, and soil microorganisms. Chem. Soc. Rev. 2021, 50, 6221-6239.
2. Zeńczak, W.; Gromadzińska, A.K. Preliminary Analysis of the Use of Solid Biofuels in a Ship's Power System. Pol. Marit. Res. 2020, 27, 67-79.
3. Başhan, V.; Demirel, H. Application of Fuzzy Dematel Technique to Assess Most Common Critical Operational Faults of Marine Boilers. Politek. Derg. 2019, 22, 545-555.
4. Luschtinetz, T.; Zeńczak, W.; Łuszczyński, D. The Selected Results of the Experimental Research of Solid Fuel Pneumatic Transportation to Ship's Boiler. Manag. Syst. Prod. Eng. 2019, 27, 144-148.
5. Taler, J.; Zima, W.; Ocłoń, P.; Grądziel, S.; Taler, D.; Cebula, A.; Jaremkiewicz, M.; Korzeń, A.; Cisek, P.; Kaczmarski, K.; et al. Mathematical model of a supercritical power boiler for simulating rapid changes in boiler thermal loading. Energy 2019, 175, 580-592.
6. Lu, X.; Liu, Z.; Ma, L.; Wang, L.; Zhou, K.; Feng, N. A robust optimization approach for optimal load dispatch of community energy hub. Appl. Energy 2020, 259, 114195.
7. Piraisoodi, T.; Iruthayarajan, M.W.; Kadhar, K.M.A. An Optimized Nonlinear Controller Design for Boiler-Turbine System Using Evolutionary Algorithms. IETE J. Res. 2018, 64, 451-462.
8. Wang, D.; Zhou, Y.; Li, X. A dynamic model used for controller design for fast cut back of coal-fired boiler-turbine plant. Energy 2018, 144, 526-534.
9. Su, Z.g.; Zhao, G.; Yang, J.; Li, Y.g. Disturbance Rejection of Nonlinear Boiler-Turbine Unit Using High-Order Sliding Mode Observer. IEEE Trans. Syst. Man Cybern. Syst. 2018, 50, 5432-5443.
10. Liu, X.; Cui, J. Economic model predictive control of boiler-turbine system. J. Process. Control. 2018, 66, 59-67.
11. Tian, Z.; Yuan, J.; Xu, L.; Zhang, X.; Wang, J. Model-based adaptive sliding mode control of the subcritical boiler-turbine system with uncertainties. ISA Trans. 2018, 79, 161-171.
12. Wei, Q.; Liu, Y.; Lu, J.; Ling, J.; Luan, Z.; Chen, M. A New Integral Critic Learning for Optimal Tracking Control with Applications to Boiler-Turbine Systems. Optim. Control Appl. Methods 2021.
13. Köhler, J.; Soloperto, R.; Müller, M.A.; Allgöwer, F. A computationally efficient robust model predictive control framework for uncertain nonlinear systems. IEEE Trans. Autom. Control 2020, 66, 794-801.

14. Reynolds, J.; Rezgui, Y.; Kwan, A.; Piriou, S. A zone-level, building energy optimisation combining an artificial neural network, a genetic algorithm, and model predictive control. Energy 2018, 151, 729-739.
15. Chi, C.; Cajo, R.; Zhao, S.; Liu, G.P.; Ionescu, C.M. Fractional Order Distributed Model Predictive Control of Fast and Strong Interacting Systems. Fractal Fract. 2022, 6, 179.
16. Ghita, M.; Cajo Diaz, R.A.; Birs, I.R.; Copot, D.; Ionescu, C.M. Ergonomic and economic office light level control. Energies 2022, 15, 734.
17. Song, D.; Chang, Q.; Zheng, S.; Yang, S.; Yang, J.; Joo, Y.H. Adaptive model predictive control for yaw system of variable-speed wind turbines. J. Mod. Power Syst. Clean Energy 2020, 9, 219-224.
18. Sorcia-Vázquez, F.; Garcia-Beltran, C.; Valencia-Palomo, G.; Brizuela-Mendoza, J.; Rumbo-Morales, J. Decentralized robust tube-based model predictive control: Application to a four-tank-system. Rev. Mex. Ingeniería Química 2020, 19, 1135 – 1151.
19. Rumbo Morales, J.Y.; López López, G.; Alvarado, V.M.; Torres Cantero, C.A.; Azcaray Rivera, H.R. Optimal Predictive Control for a Pressure Oscillation Adsorption Process for Producing Bioethanol. Comput. Sist. 2019, 23, 1593-1617.
20. Morales, J.Y.R.; López, G.L.; Martínez, V.M.A.; Vázquez, F.d.J.S.; Mendoza, J.A.B.; García, M.M. Parametric study and control of a pressure swing adsorption process to separate the water-ethanol mixture under disturbances. Sep. Purif. Technol. 2020, 236, 116214.
21. Zhang, X.; Zhang, L.; Zhang, Y. Model predictive current control for PMSM drives with parameter robustness improvement. IEEE Trans. Power Electron. 2018, 34, 1645-1657.
22. Shen, C.; Shi, Y.; Buckham, B. Trajectory tracking control of an autonomous underwater vehicle using Lyapunov-based model predictive control. IEEE Trans. Ind. Electron. 2017, 65, 5796-5805.
23. Zhu, H.; Zhao, G.; Sun, L.; Lee, K.Y. Nonlinear predictive control for a boiler-turbine unit based on a local model network and immune genetic algorithm. Sustainability 2019, 11, 5102.
24. Zhang, Y.; Decardi-Nelson, B.; Liu, J.; Shen, J.; Liu, J. Zone economic model predictive control of a coal-fired boiler-turbine generating system. Chem. Eng. Res. Des. 2020, 153, 246-256.
25. Liu, X.; Cui, J. Fuzzy economic model predictive control for thermal power plant. IET Control Theory Appl. 2019, 13, 1113-1120.
26. Wang, G.; Wu, J.; Ma, X. A nonlinear state-feedback state-feedforward tracking control strategy for a boiler-turbine unit. Asian J. Control 2020, 22, 2004-2016.
27. Kong, L.; Yuan, J. Disturbance-observer-based fuzzy model predictive control for nonlinear processes with disturbances and input constraints. ISA Trans. 2019, 90, 74-88.
28. Cui, J.; Chai, T.; Liu, X. Deep-neural-network-based economic model predictive control for ultrasupercritical power plant. IEEE Trans. Ind. Inform. 2020, 16, 5905-5913.
29. Sanchez, R.O.; Rumbo Morales, J.Y.; Ortiz Torres, G.; Pérez Vidal, A.F.; Valdez Resendiz, J.E.; Sorcia Vázquez, F.d.J.; Nava, N.V. Discrete State-Feedback Control Design with D-Stability and Genetic Algorithm for LED Driver Using a Buck Converter. Int. Trans. Electr. Energy Syst. 2022, 2022, 8165149.

30. Nedić, A.; Liu, J. Distributed optimization for control. Annu. Rev. Control Robot. Auton. Syst. 2018, 1, 77-103.
31. Bell, R.; Åström, K.J. Dynamic Models for Boiler-Turbine-Alternator Units: Data Logs and Parameter Estimation for a 160 MW Unit; Technical Reports; Lund Institute of Technology: Lund, Sweden, 1987; TRFT-3192.
32. De Keyser, R. Model based predictive control for linear systems. In UNESCO Encyclopaedia of Life Support Systems, Robotics and Automation; Article Contribution 6.43.16.1; Eolss Publishers Co., Ltd.: Oxford, UK, 2003; Volume XI.
33. De Keyser, R.; Ionescu, C.M. The disturbance model in model based predictive control. In Proceedings of the Proceedings of 2003 IEEE Conference on Control Applications, Istanbul, Turkey, 25 June 2003; IEEE: Piscataway, NJ, USA, 2003; Volume 1, pp. 446 – 451.
34. Fernandez, E.; Ipanaque, W.; Cajo, R.; De Keyser, R. Classical and advanced control methods applied to an anaerobic digestion reactor model. In Proceedings of the 2019 IEEE CHILEAN Conference on Electrical, Electronics Engineering, Information and Communication Technologies (CHILECON), Valparaiso, Chile, 13-27 November 2019; IEEE: Piscataway, NJ, USA, 2019 ; pp. 1-7.
35. Romero, M.; de Madrid, A.P.; Vinagre, B.M. Arbitrary real-order cost functions for signals and systems. Signal Process. 2011, 91, 372-378.
36. Zhao, S.; Cajo, R.; De Keyser, R.; Liu, S.; Ionescu, C.M. Nonlinear predictive control applied to steam/water loop in large scale ships. IFAC-PapersOnLine 2019, 52, 868-873.

CHAPTER
10

Effect of Waste Heat Utilization on the Efficiency of Marine Main Boilers

Cezary Behrendt and Marcin Szczepanek

Faculty of Marine Engineering, Maritime University of Szczecin, ul. Waly Chrobrego 1-2, 70-500 Szczecin, Poland

ABSTRACT

As a result of technological advancements in the 21st century, steam-driven ships, with an efficiency of between 36 and 38%, have been replaced by diesel-driven ships, with an efficiency of up to 55%. Accordingly, the manufacturers of the main boilers and steam turbines in ships have fought for their products to remain on the market by modernizing their construction in order to increase their competitiveness in relation to diesel drives. Based on a review of data in the literature, this article presents an analysis of the increased efficiency of ship boilers with the use of waste heat from boiler exhaust gas. This paper considers the utility of using this heat to heat up the water that supplies the boilers as well as the use of the heated air in the combustion process. The analysis takes into account the different types of heat exchangers used for these purposes and the different efficiencies of boilers made by the leading boiler manufacturers. The formulae and calculation methodologies used are applied in the following analysis of the impact of using waste heat to increase boiler efficiency.

Keywords: *marine main boilers; waste heat recovery; feeding water heating; air heating; boilers efficiency*

1. INTRODUCTION

Turbo steam power plants were popular in 1960-2005 on ships of high load bearing capacity and high demand for drive power in the range between 20 and 50 MW (tankers, passenger ships, LNG tankers). Such powers were achievable only by steam turbines due to the low power of diesel engines produced at that time. The development of ship construction of diesel engines by increasing their power, introducing dual fuel engines, as well as using systems for the deep utilization of waste heat [1], made it possible for the diesel power plants to work more efficiently, up to 50%. This resulted in decreasing the number of ships with steam drive being built.

It is worthwhile to note that until 2005 all LNG tankers were equipped with a turbo steam drive, which is the only one that makes it possible to use naturally evaporating gas as a fuel in main boilers. The main drawback of turbo steam power plants is their low efficiency, in the range of 36-38%. Increasing the efficiency of ship turbo steam power plants resulted in increasing the parameters of steam produced in the main boilers. In the years 1995-2010, the pressure of steam produced in ships' main boilers was ranging between 6.0 and 6.8MPa, and steam's superheating temperature was $515-525°C$[2, 3]. The improvement of material engineering allowed for the construction of experimental ship turbo steam power plants where main boilers produced steam with a pressure of 10-12 MPa, and a superheating temperature of 560°C [4-6]. Apart from increasing the values of steam parameters, using steam inter-stage superheating, carnotization, and the use of waste heat, new boiler designs have been introduced to increase the speed of natural circulation [2]. These procedures allowed for achieving the efficiency of ship turbo steam power plants in the range of $39-41\%$ [7].

The article presents an analysis of the methods and impact of the use of waste heat on the efficiency of the ship's main boilers, based on the literature, the ship's technical documentation, and the results of operational tests. The literature mainly deals with the development of boiler designs, increasing steam parameters and materials used for their construction in order to increase the efficiency of marine turbo steam power plants. The advisability of using waste heat in exhaust gases to heat boiler feed water and air to raise combustion temperature has also been discussed [2-4,8-10]. Boiler manufacturers' company materials [11-13] and vessel technical documentation were used to obtain data on the achieved temperatures of heated air supplied to the boiler furnace chamber. The methodology for determining the effect of exhaust gas heat recovery on exhaust loss and the effect of air temperature on combustion temperature was developed based on [5, 6, 14, 15].

The main boilers of the following companies were taken into account: Mitsubishi, Kawasaki, Greens Power, and Combustion Engineering.

2. PURPOSEFULNESS OF USING WASTE HEAT FROM THE EXHAUST EMISSIONS

The greatest resource of waste heat is included in exhaust emissions coming out from the boiler, whose temperature affects the boiler's outlet loss value. On the basis of calculation methods presented in the literature [4, 9] we determined the values of outlet loss depending on exhaust emissions temperature.

The empirical Formula (1) in the literature [4, 9] was used to calculate the outlet loss $S_{out;}$;

$$S_{out} = \left[\left(\frac{0.233}{CO_2} - 0.0053\right)\left(\frac{131x10^3}{W_d} - 1\right)\right](t_{out} - t_{env}) \tag{1}$$

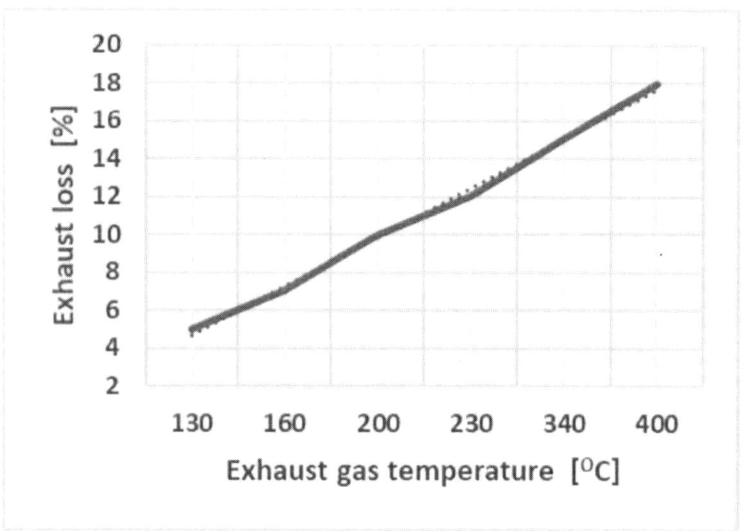

Figure 1. *Relation between exhaust gas temperature and boiler's exhaust loss [own study].*

where:
t_{out} [°C]-exhaust gas temperature at the outlet from the boiler;
t_{env} [°C]-ambient temperature;
CO_2[%]-carbon dioxide content in the exhaust gas;
W_d $\left[kJ/kg, kJ/um^3 \right]$-fuel calorific value;

The calculations were performed after assuming the constant values of t_{env}, CO_2, and W_d, so the calculated values of the outlet loss S_{out} depended only on the changes in the temperature tout of the exhaust gases leaving any main boiler. The calculations were made on the assumption that the flue gas temperature tout was changed at the boiler inlet in the range from 130°C to 400°C.

The calculations' results allowed for drawing Figure 1, which presents outlet loss values depending on exhaust gas temperature.

The highest calculated value of the exhaust loss was 18% for an exhaust gas temperature of 400°C (Figure 1). This loss value occurs when the waste heat resou rces of the flue gas are not used. The minimum value of this loss of 5% occurs with the maximum use of waste heat while cooling the flue gas to a temperature of 130°C, which does not cause low-temperature corrosion yet [2-4].

In order to recover waste heat, additional heat exchangers are installed in the boiler flue gas outlet system in the form of feed water heaters and air heaters. The use of waste heat will result in cooling down the flue gas to the temperature acceptable from the point of view of low-temperature corrosion (130 – 150°C), which will allow it to obtain the outlet loss value in the range of 5 – 6% (Figure 1).

Figure 2 presents examples of installations of additional heat exchangers in main boilers.

In the case of MB boilers made by Mitsubishi, two methods of exhaust gas waste heat utilization have been proposed. The first method (Figure 2a) includes

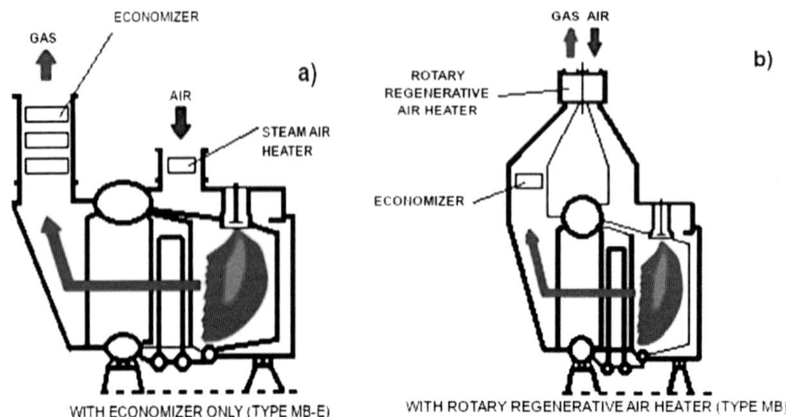

Figure 2. *Options of waste heat utilization for MB boiler of Mitsubishi: (a) multistage water heater, (b) water and air heater* [4,6].

using waste heat to heat water feeding the boiler in a multistage water heater. For the purpose of heating the air, a heat exchanger fed with saturated steam of the pressure from 0.6 to 2.0MPa were assembled [5,7]. The second method (Figure 2b) consists in the dividing the waste heat for the purpose of heating both the water feeding the boiler and the air needed in the combustion process.

The choice of the utilization method of the waste heat of the boiler exhaust gas depends on the applied other methods of increasing the efficiency of the turbo steam power plant. In the case when carnotization is used in the engine room (heating of the feed water with the use of bleed steam from turbines), the boiler feed water temperature reaches high values (in the range 190 – 230°C, see Table 1). One water heater is used to heat it up to the saturation temperature for the pressure and combustion air heater (Figure 2b). If no carnotization is used in the turbo steam power plant, the temperature of the boiler feed water is 110 – 140°C, and in order to heat it up to the saturation temperature, more heat is required, which requires the use of multi-stage feed water heaters. Meanwhile, for air heating heat exchangers supplied with saturated steam are used (Figure 2a).

3. FEED WATER HEATING

The heat exchange process in a steam boiler consists of three stages: water heating from the supply temperature to the saturation temperature for a given operating pressure, water evaporation, and steam superheating. Based on the literature [1,9], the percentage of heat required to heat water to the saturation temperature was determined and presented in Figure 3 in relation to the total heat required for the above-mentioned three stages of the heat exchange process.

As can be seen in Figure 3, in the ship's main boilers, where operating pressures in the range of 6.3-12.0 MPa are used, 28 – 40% of the total amount of heat generated during fuel combustion is used to heat water to the saturation temperature in the boiler's combustion chamber. The use of a combustion water heater

Table 1. *Comparison of the method of utilizing exhaust gas waste heat from the main boilers* [3 – 5, 8, 9, 12 – 14, 16 – 19].

Producer	Boiler Type	Working Pressure	Steam Temp.	Range of Nominal Capacities	Feedwater Temp.	Air Temp.	Operational Efficiency (at HFO)
		MPa	°C	t/h	°C	°C	%
Mitsubishi	MB	6.15	515	23–75	210	150	90.0
	MB-E	6.15	515	15–70	138	120	88.5
	MBR	10.0	560	40–70	138	120	88.5
Combustion Eng.	V2M8	6.2	515	30–85	183	290	89.4
	V2M9	6.6	515	60–140	190	200	90.0
Green Power	ESD-IV	6.2	515	36–65	180	165	91.3
Kawasaki	UM	6.1	525	47–143	205	155	90.0
	UME	6.1	525	47–143	145	130	88.5
	UFR	10.3	525	50–140	195	130	90.2
	UTR-II	12	565	35–100	229	133	90.2

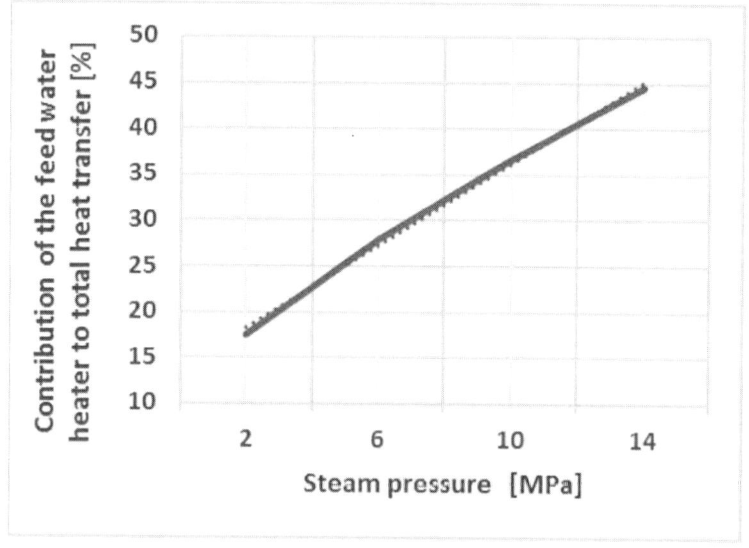

Figure 3. *Percentage share of total heat taken over by the feed water heater depending on the operating pressure at the constant temperature of feed water and steam temperature [own study].*

will significantly reduce this amount of heat or bring it to zero when the temperature of water heating reaches the saturation temperature. As a result, the amount of fuel burned is reduced while maintaining a constant amount and parameters of the steam produced, or increasing the amount of steam produced while maintaining a constant amount of fuel burned. In both cases, there is also an increase in boiler efficiency.

Theoretically, in combustion water heaters supplying boilers, the water should be heated to the saturation temperature, namely 280 – 324°C for the relevant working pressures of 6.3-12.0 MPa. In practice, depending on the amount of waste heat contained in the exhaust emissions, two types of water heaters are used: - Water heaters without water evaporation-the water is heated to a temperature of 15-30 °C

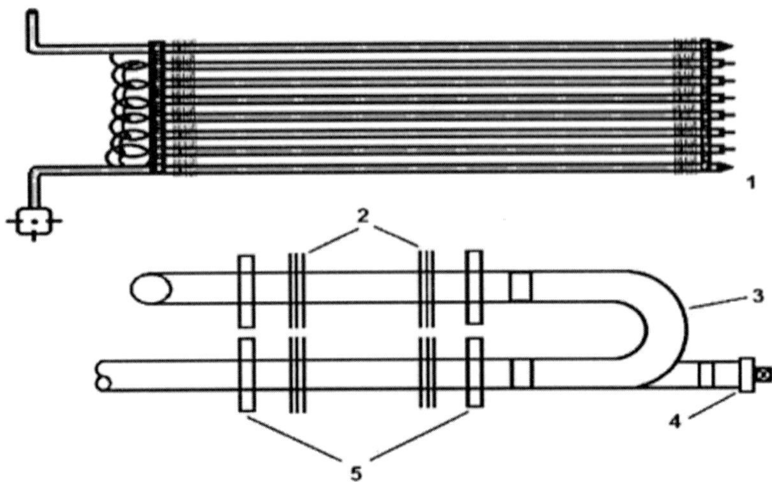

Figure 4. *Feed water boiler heater [7]. 1-steel pipe; 2-fins; 3-elbow; 4-cleanout; 5-flange.*

lower than the saturation temperature. Their construction is based on a bundle of steel pipes, often finned, rolled into coils, and welded at their ends into collectors. The whole as one assembly of modular structure is placed in the chimney chute. The modular structure facilitates access to the heater during renovation works.

- Water heaters with water evaporation-the water is heated to the boiling point and it partially evaporates. The degree of wet steam dryness is low and practically amounts to $x = 0.02 \div 0.10$[7, 16]. Higher degrees of dryness are not used, because when $x \geq 0.15$, there will be a sharp increase in resistance and flow disturbances. Therefore, these types of heaters require precise hydrodynamic calculations.

The diagram of the feed water boiler heater Is shown in Figure 4, while the heater module is shown in Figure 5.

4. AIR HEATING

The heated air will bring additional heat into the combustion chamber. This results in an increase in the theoretical combustion and flue gas temperature and an increase in flue gas enthalpy.

Based on the methodologies of calculating the combustion temperature in the boiler presented in the literature [9], the increase in combustion temperature depen ding on the air temperature was calculated. The results of the calculations allowed for drawing Figure 6.

On the basis of Figure 6, it can be inferred that air temperature rising by 100°C makes combustion temperature increase by about 70°C.

Air temperature rising will also make the fuel injected into the combustion chamber evaporate faster. It also makes the combustion process itself take place faster and improves its quality.

Figure 5. *Water heater module of Degetron Engineering [17].*

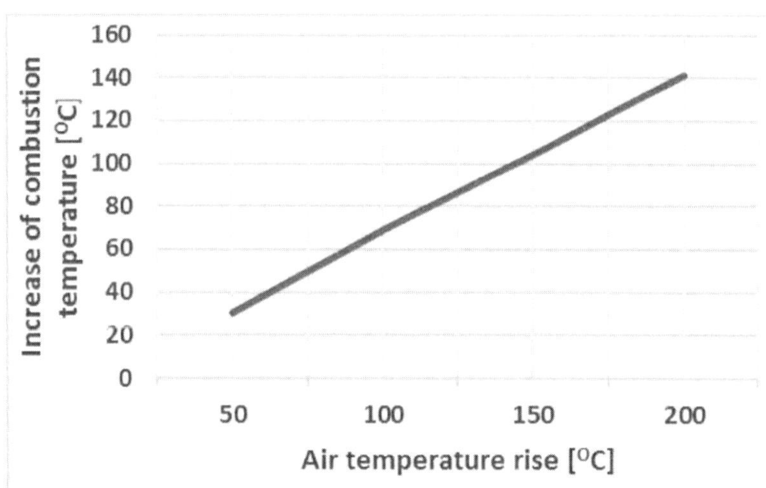

Figure 6. *Increase of combustion temperature depending on air temperature [own study].*

Apart from pipe, usually spiked and finned air heaters and also rotary air heaters (regenerative air heaters) are often used-Figure 7.

The main elements of this type of heater are (Figure 7): a rotating drum with an installed package of thin corrugated sheets with a thickness of about 0.5 mm, and an electric motor which, through a mechanical transmission, gives the drum a rotational speed in the range of 1-2 rev/min. Half of the sheet packet is in the flue gas stream receiving heat from it, and then, due to the rotational movement, the sheet packet is moved into the air flow channel, where the air washing around the

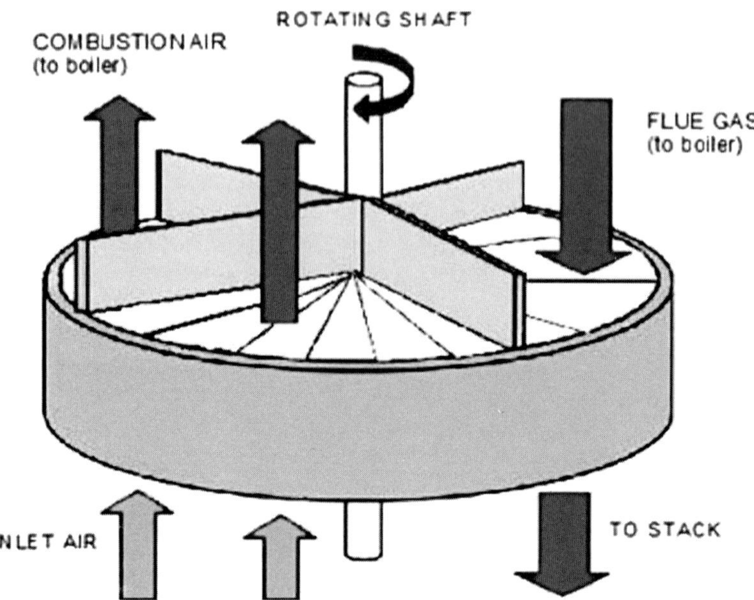

Figure 7. *Rotary air heater (own drawing based on [14]).*

sheet packet is heated. One of the most important operational advantages of rotary heaters is their high resistance to low-temperature corrosion. When burning fuels with a high sulphur content of 3-3.5%, it was observed that they were practically non-corrosive [7]. Another important advantage is the self-cleaning of the heater. The heat exchange surfaces are washed alternately with exhaust gases and air of different velocities, flow directions, and temperatures. These factors make it much harder for contaminants to settle.

5. SUMMARY

In all main boilers installed on the ships with turbo steam power plants, exhaust waste heat is used for heating feed water and the air necessary for the combustion process. In order to assess the temperature of the heated air, temperature of feed water before the diesel heater, as well as boilers' efficiency, Table 2 has been drawn.

Table 2 presents a list of heat exchangers for flue gas waste heat recovery, in the main boiler types most frequently used in the fleet.

6. CONCLUSIONS

The standard parameters of steam produced in modern main boilers are a working pressure of 6.0-6.6 MPa and a steam superheating temperature of 510 – 515°C. The manufacturers of main boilers strive to maximize these parameters and use secondary steam superheating in order to increase the efficiency of turbo steam power plants. As a result, the MBR and UTR-II boilers (operating pressure 10.0-12.0 MPa, steam superheating temperature 560-565 °C) have been introduced to

Table 2. *Methods and heat exchangers for the utilization of waste heat from the main boilers* [3 – 5,7,8,10,14 – 19].

	Boiler Type							
Element of the Waste Heat Recovery system	Mitsubishi MB	Mitsubishi MBR	CE * V2M8	CE * V2M9	Greens Power ESD IV	Kawasaki UM	Kawasaki UFR	Kawasaki UTR-II
One-section feed water heater			Option					
Multi-sectional feed water heater	MB-E					UME		
Rotary air heater								
Heating the air from the boiler jacket								
Support the heating of the feed water with additional coil sections								
Steam air heater	MB-E							

* Combustion Engineering.

the market. This is primarily due to the high market competitiveness of diesel combustion engines (mainly dual-fuel) for ship propulsion systems.

Modern main boilers are characterized by very high efficiency, exceeding 88%. Such high efficiencies are achieved by the utilization of waste heat for heating the feed water and air. A barrier to further utilization of heat contained in exhaust gas is the threat of aggressive exhaust components reaching the dew point and the occurrence of lowtemperature corrosion. The highest recorded efficiencies are in the range between 90.0 and 91.3% (Table 2). The ESD IV boiler additionally uses the external boiler walls to preheat the air, thereby reducing the amount of heat radiating outside and allowing achievement of the highest efficiency of 91.3%. The lowest efficiencies are observed for the following types of boilers: MB-E, MBR, and UME, which are equipped with multi-section feed water heaters only. At nominal loads, the design of these heaters does not allow the exhaust gases to cool below a temperature of 170°C. Therefore, the efficiency of this group of boilers is 88.5%. Boiler manufacturers provide their efficiency values at nominal efficiency and when using 100% liquid fuel with a gross calorific value (Wd = 42.500 kJ/kg).

The manufacturers of the main boilers report the feed water temperature at the inlet to the internal boiler heater. For boilers equipped with multi-section feed water heaters, the inlet water temperature is quite low and amounts to about 1400° C (e.g., MB-E, UME, and MBR boilers). Boilers equipped with a single-section heat er require preliminary heating of the water to a temperature of about 200°C (e.g., MB, UM, and V2M9 boilers).

The heaters constructed without an internal boiler feed water heater need to heat the water to the highest possible temperature in the exchangers supplied by steam from the main drive turbines. Usually, the water is heated to a temperature of about 200°C (e.g., V2M8 type boiler).

The temperature of the heated air supplying the furnace chamber, as well as the water temperature, is very diverse and depends mainly on the structure of the waste heat utilization section. The main boilers that use steam heaters to heat the air are able to increase the temperature to about 130°C (these are boilers equipped with a multi-section water heater which reduces the exhaust gas temperature at the outlet to a minimum, i.e., MB-E, UME, and MBR types). Boilers equipped with a rotary air heater increase the temperature to about 150°C (cooperating with a single-section water heater, e.g., UM boiler). The design of the waste heat utilization section of the V2M8 boiler consists solely of a rotary air heater. This allows the air to be heated up to 290°C (with a nominal amount of steam produced).

Author Contributions

Author Contributions: Conceptualization, C.B. and M.S.; methodology, C.B. and M.S.; validation, C.B. and M.S.; formal analysis, C.B. and M.S.; investigation, C.B. and M.S.; resources data curation, C.B. and M.S..; writing-original draft preparation C.B.; writing-review and editing, C.B. and M.S.; visualization, M.S.; supervision, C.B.; project administration C.B.; funding acquisition, C.B. and M.S. All authors have read and agreed to the published version of the manuscript.

Acknowledgments

This research outcome has been achieved under the research project no. 2/S/ KSO/ 2022 financed from a subsidy of the Polish Ministry of Science and Higher Educations for statutory activities of Maritime University of Szczecin.

REFERENCES

1. Behrendt, C. Conditions of Marine Waste Heat Recovery in Marine Waste Heat Recovery Systems. J. Mach. Constr. Maint. 2017, 107, 141 – 147.
2. Behrendt, C. Development of Main Marine Boilers' Structures. In Proceedings of the GEOLINKS International Conference on GeoSciences 2019, Athens, Greece, 26-29 March 2019; Volume 1, pp. 127-136. [CrossRef]
3. Górski, Z.; Perepeczko, A. Okrętowe Kotty Parowe; Fundacja Rozwoju Wyższej Szkoły Morskiej w Gdyni: Gdynia, Poland, 2002; ISBN 83-87438-37-5.
4. Goszczyński, P. Analiza Rozwiązań Konstrukcyjnych Okrętowych Kotłów Głównych (Analysis of Main Marine Boilers' Structures). Master's Thesis, Maritime University of Szczecin, Szczecin, Poland, 2013.
5. Kitto, J.B.; Stultz, S.C. Steam-It's Generation and Use; Babcock & Wilcox, McDermott 2Company: Barberton, OH, USA, 2005 .
6. Yi, Y.-S.; Watanabe, Y.; Kondo, T.; Kimura, H.; Sato, M. Oxidation Rate of Advanced Heat-Resistant Steels for Ultra-Supercritical Boilers in Pressurized Superheated Steam. J. Press. Vessel. Technol. 2001, 123, 86-98. [CrossRef]
7. Ito, M.; Hiraoka, K.; Matsumoto, S.; Tsumura, K. Development of High Efficiency Marine Propulsion Plant (Ultra Steam Turbine). 2007. Available online: http://www.mhi.co.jp/technology/review/pdf/e443/e443015.pdf (accessed on 8 September 2022).
8. Kruczek, S. Kotly, Konstrukcje i Obliczenia; Oficyna Wydawnicza Politechniki Wrocławskiej: Wroclaw, Poland, 2001; ISBN 978-83-7085-600-7.
9. Rokicki., H. Urzadzenia Kotłowe. Przykłady Obliczeniowe; Gdansk Technical University: Gdańsk, Poland, 1996; ISBN 9788386537341.

10. Nakase, T.; Uchida, H.; Yamaguchi, D.; Itoh, T. Challenge to 12MPaG Steam Pressure UTR-II Type Marine Reheat Boiler for LNG Carriers; Kawasaki Heavy Industries Ship Engineering: Tokyo, Japan, 2008; pp. 12- 19. Available online: http:/ / global.kawasaki.com/en/corp/rd/ magazine/166/index.html (accessed on 12 September 2022).
11. Available online: http://global.kawasaki.com/en/energy/solutions/energy_plants/marine.html (accessed on 23 September 2022).
12. LNGC Machinery System Operation Manual 2007. Available online: https://www.scribd.com/doc/105393590/LNGCMachinery-System-Operation-Manual (accessed on 21 November 2022).
13. British Endurance Machinery Manual. Available online: https://www.scribd.com/document/530159691/British-EnduranceMachinery-Manual (accessed on 21 November 2022).
14. Regenerative Air Pre-Heater. Available online: https://www.lnthowden.com/product-services/regenerative-air-pre-heater / (accessed on 11 September 2022).
15. Available online: https://heatexchangegroup.co.uk/wp-content/uploads/2019/07/Greens-marine-boiler-services-online.pdf (accessed on 23 September 2022).
16. Muślewski, Ł.; Pajakk, M.; Landowski, B.; Żółtowski, B. A Method for Determining the Usability Potential of Ship Steam Boilers. Pol. Marit. Res. 2016, 23, 105-111.
17. Available online:http://www.trade.indiamart.com(accessed on 21 September 2022).
18. Winkler, W. Design of Modern Marine Boilers; Babcock & Wilcox, McDermott Company: Barberton, OH, USA, 1998 .
19. Josefsson, L. Combustion Engineering V2M8 Boiler. 2021. Available online: http://steamesteem.com/?boilers/combustionengineering-v2m8.html (accessed on 22 September 2022).

CHAPTER 11

Assessing the Macroeconomic and Xocial Impacts of Slow Steaming in Shipping: A Literature Review on Small Island Developing States and Least Developed Countries

Seyedvahid Vakili, Fabio Ballini, Alessandro Schönborn, Anastasia Christodoulou, Dimitrios Dalaklis and Aykut I. Ölçer

World Maritime University, Malmö, Sweden

ABSTRACT

The International Maritime Organisation (IMO) has adopted the Energy Efficiency Existing Ship Index (EEXI) and the Carbon Intensity Indicator (CII) as short term measures for decarbonisation of the shipping industry; the IMO also made the collection of relevant data and associated reporting of the indicator mandatory from January 2023. However, many existing ships do not meet the EEXI and CII "targets" and cannot invest in other technologies to meet the relevant requirements. Given the various barriers to energy efficiency, the application of slow steaming may be a measure to effectively meet EEXI and CII *requirements. A qualitative systematic literature review was conducted on the potential macroeconomic and social impacts of slow steaming on states, with a special focus on Small Island Development States and Least Developed Countries, when used as the primary modality of reducing GHG emissions from shipping. This effort includes peer-reviewed studies and studies from the gray literature, many of which include examples that borrow data from the aftermath of the economic crisis that was manifested in 2008. The vast majority of those studies is focused on the economic cost-effectiveness or impact on transportation costs when using slow-steaming as a means of reducing marine fuel consumption. Moreover, a number of these studies were relying on modeling techniques, by using a limited number of ships and associated routes to determine the effects of slow-steaming. A reasonable degree of agreement emerged from the literature that a reduction in transportation costs results from a reduction in speed, being attributed primarily to reduced fuel costs, with which it is associated. Other cost-increasing factors, such as vessel operating costs, had a less dominant effect. The literature often pointed out that the cost reduction resulting from the application of slow-steaming was unevenly distributed among maritime stakeholders. Shipping companies were the main beneficiaries of significant cost savings, but these "savings" were not always passed on to shippers.*

Keywords: *Air emissions; Decarbonisation of international shipping; IMO's short term measure; Least Developed Countries; Slow steaming; Small Island Development States; Sustainable shipping; Transportation cost*

1. INTRODUCTION

Maritime transportation is associated with a contribution in carriage of 80% of cargo by volume and 70% by value; it is clearly the backbone of global trade (UNCTAD 2021). Although the shipping industry plays a crucial role in trade and supply chain as well as creating jobs, and economic progress for societies, it also clearly associated with certain negative externalities. Maritime transport was responsible for 2.89% of global Greenhouse Gas (GHG) emission in 2018, and its general trend since 2013/2014 has been for increasing GHG from shipping (9.6%) and international shipping (5.6%) (Fourth IMO GHG Study 2020). On the positive side, at the 72 nd session of Marine Environment Protection (MEPC), the IMO (MEPC 2018) adopted the Initial IMO Strategy on reduction of GHG emissions from ships (Initial Strategy) in 2018, with a vision to phase out GHG emissions from international shipping, as soon as possible in the twenty-first century. The ambition of this strategy is to reduce average carbon dioxide (CO_2) emissions per transport work by 40% by 2030 , and by 70% by 2050 compared to 2008 . The total annual GHG emissions related to international shipping are aimed to be reduced by 50% by 2050 , and to reach zero before the end of the century (Vakili et al. 2022a, b, c).

Current measures addressing GHG emissions are regulated under International Convention for the Prevention of Pollution from Ships (MARPOL) Annex VI and are comprised by the Energy Efficiency Design Index (EEDI), the Ship Energy Efficiency Management Plan (SEEMP), and the Fuel Oil Consumption Data Collection System (DCS), mandating annual reporting of CO_2 emissions. The IMO has adopted a two-tier approach to implementing decarbonisation measures, focusing first on a limited set of short-term measures, before implementing more comprehensive medium- and longterm measures. At the Marine Environment Protection Committee (MEPC) 75 meeting (MEPC 2020) a combined short-term measure was agreed, comprising a technical measure (Energy Efficiency Existing Ship Index, EEXI) and an operational measure (Carbon Intensity Indicator, CII) (Vakili et al. 2021). This approved short-term measure defines a carbon intensity reduction goal, but it does not specify the means by which it will be achieved. Amongst the guiding principles of the Initial Strategy is the need for the impacts of the measures on States to be assessed, with particular attention to developing countries, especially small island development States (SIDS) and least developed countries (LDCs) (MEPC 304(72)).

The aim and the originality of this paper is to identify the potential impacts of slow steaming, when used as a short-term measure. MEPC.1/Circ.885 (MEPC

2019), provides a description of the procedure for the impact assessment of the proposed measures. The impact assessment must be simple, inclusive, transparent, flexible, evidence-based and measure-specific, as well as proportionate to the complexity and nature of the candidate measures in alignment with consideration of the candidate measures development. Additionally, the initial impact assessment must pay particular attention to the needs of developing countries, especially Small Island developing States (SIDS) and least developed countries (LDCs). Moreover, the impact assessment should consider, as appropriate, inter alia (1) geographic remoteness of and connectivity to main markets; (2) cargo value and type; (3) transport dependency; (4) transport costs; (5) food security; (6) disaster response; (7) cost-effectiveness; and (8) socio-economic progress and development. The above criteria formed the basis for the analysis of those articles identified in the relevant literature review. As already pointed out, the goal-based measure involves a limitation of operational emissions, which means that the speed and other operational characteristics must follow the carbon intensity target in practice. Although slow steaming is not equivalent to the goal-based measure, the target-based measure can lead to speed reductions. For existing vessels, the most obvious operational measure that can be used to meet the operational goal-based objectives is a reduction in speed, and a similar situation applies to technical measures, since EEXI or power reduction would lead to a reduction in speed, and the owner, operator or charterer of the vessel is the relevant decision-maker (Vakili et al. 2022a).

The initial impact assessment in this research effort seeks to identify both positive and negative potential impacts of slow-moving traffic and to analyse the magnitude of the effects on transport costs, trade and those relating to Gross Domestic Product (GDP). In addition, the initial impact assessment seeks to evaluate whether the measure is likely to lead to disproportionately negative impacts and, if so, how they can be managed (e.g., avoided, addressed or mitigated), if appropriate.

This document is structured as follows: "Introduction" section provides the necessary Introduction; "Methodology" section describes the methodology of the study that was based on a systematic literature review; "The analysis of the peer-reviewed and grey literature" section summarises the analysis of the peer-reviewed and grey literature, and discussions and conclusions are presented in "Discussion" and "Conclusions" sections respectively.

2. METHODOLOGY

This qualitative systematic literature review aimed at identifying the potential impact of short-term measures of the IMO's Initial Strategy-more specifically slow steaming- on States, as reported in the literature. Particular attention was paid to the needs of developing countries, especially SIDS and LDCs, with regard to the eight impact criteria provided in MEPC.1/Circ.885 (MEPC 2019). This literature review process includes relevant peer-reviewed articles, and various relevant grey literature (non-peer reviewed) documents. A systematic process was used to collect peer reviewed literature, as well as grey literature, in order to provide an objective means of identifying the sources.

Table 1. *Inclusion and exclusion criteria*

Criterion one (Filtering step 1)	Criterion two (Filtering step 2)	Criterion three (Filtering step 3)
Peer reviewed articles and high quality conference papers, books, industrial and technical reports, which are relevant to research questions and address slow steaming within the eight impact criteria provided in MEPC.1/Circ.885	Non-English studies, duplicate articles, non-peer reviewed articles, low quality industrial and technical reports, and articles that not totally covered the topic	Exclude texts older than 2003

The literature was identified by keyword searches in the following abstract and citations databases: Scopus, Science Direct, Google Scholar, Research Gate, and EBSCO. The specific keywords and their combinations used were as follows: "slow steaming", "speed reduction", "impact assessment", "engine power limit", "shaft power limit", "Developing Countries", "SIDs", "LDCs", "connectivity", "geo graphic remoteness", "cargo value", "cargo type", "transport dependency", "transport cost", "freight", "inventory cost", "food security", "disaster response", "socio-economic progress and development", "logistic" and "cargo quality".

A total of 864 works in eight categories were initially identified. To select the most relevant articles, two criteria for inclusion and exclusion were considered (see Table 1). In the first filtering step, where titles, conclusions, abstracts and keywords were analysed and considered, 469 studies were excluded and only 395 articles that were found to be relevant according to exclusion criterion one (Table 1). In the Table 1 second filtering step, the exclusion criterion two (Table 1) was applied after reading and analysing the whole text. In this second step, 315 articles were found not to explicitly contribute to the topic, and were then excluded. In addition, 15 articles were included through backward and forward snowballing, which resulted in a total of 95 documents. In a third filtering step, any works older than 2003 were excluded, leaving 62 articles in the final selection. With the help of figure one, the various filtering steps that served the methodology are summarised (Fig. 1).

3. THE ANALYSIS OF THE PEER-REVIEWED AND GREY LITERATURE

This section contains the analysis of the peer-reviewed literature, along with the identified relevant grey literature. Findings are grouped with regards to each impact criterion by referring to a selection of relevant findings from the wider literature portfolio. The literature reviewed showed a stark variation in the frequency with which the literature treated the various impact criteria (Fig. 2 and Table 2). It was found that the majority of studies addressed the impact of slow-steaming on transport dependency and transport costs. A high number of studies considered cargo-value and type when discussing potential impacts. An intermediate number of studies discussed cost-effectiveness, socio-economic progress and development, and geographic remoteness and connectivity to main markets. It is also interesting to note that very few studies addressed the issues of food security and disaster response. The frequency with which the impact criteria were observed in the literature, was a mere observation of this study, and to some extent indicative of gaps in the literature. It is by no means representative of a relative importance amongst any of the impact criteria.

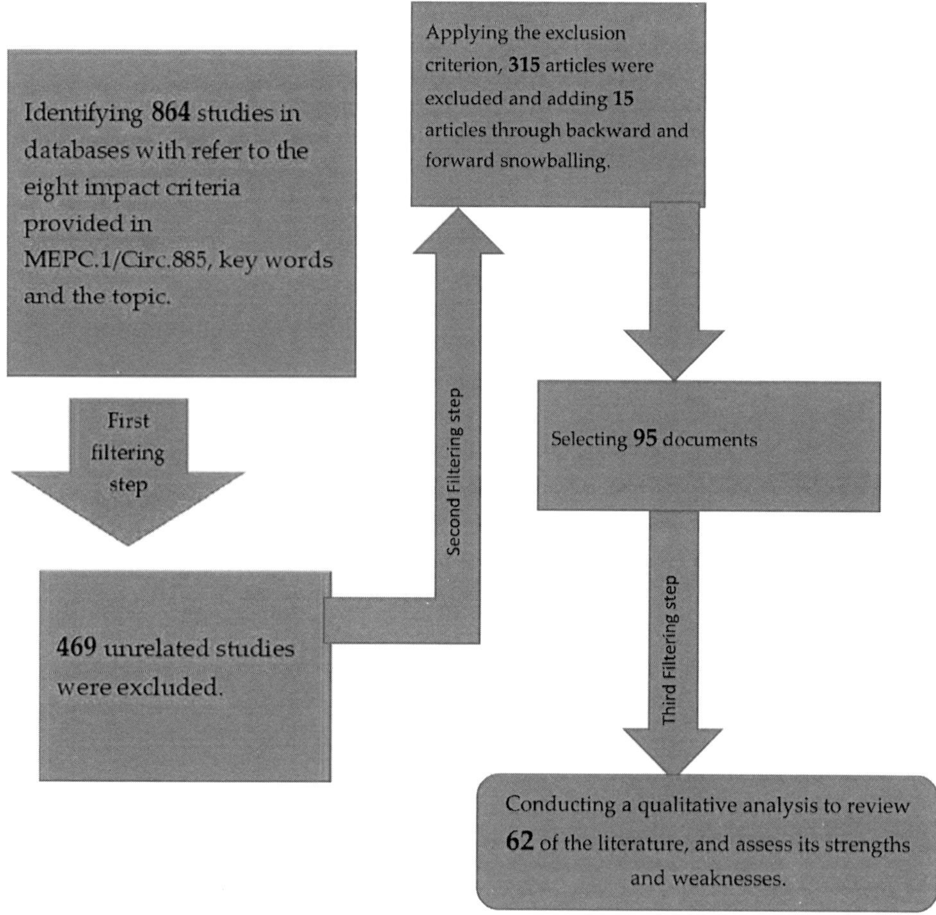

Figure 1. *The flow chart of the methodology*

It is important to highlight that a comprehensive impact assessment of all measures depends on the complexity and characteristics of the measures under examination. While it may be possible to assess the impact of measures against these eight criteria, it is difficult to identify all negative impacts and disproportionate impacts on developing countries, SIDS and LDCs. As the eight criteria are interlinked and interrelated for a more accurate and comprehensive impact assessment of measures, it is important to further define the eight impact criteria, such as transport dependence and others, and consider their interlinkages.

3.1. Geographic remoteness of and connectivity to main markets

Several peer-reviewed articles have examined the relation of slow steaming to geographically remote and distant areas from main markets. Hummels and Schaur (2012), as well as Maloni et al. (2013), discussed the problem of increased "pipeline inventory" (i.e. goods in transit between buyer and seller), including increased

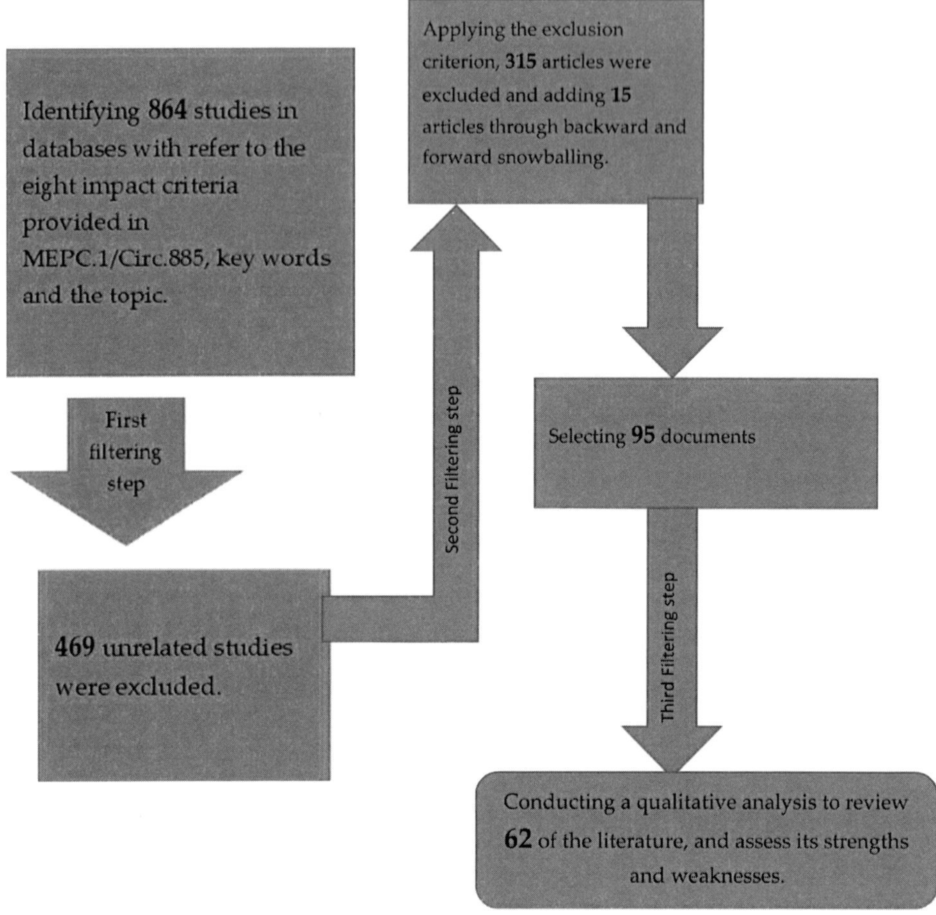

Figure 2. *Frequency of the various impact criteria in the examined literature*

transport costs due to slow steaming for areas distant to main markets. Additionally, a study of Wilmsmeier and Sanchez (2009) showed that food prices, which are correlated with transport costs, are higher in more geographically remote areas, not only by geographical distance, but also by their position in global shipping networks, for example if they are situated at the start or end of shipping lines. Another important aspect discussed in this research effort is that it is necessary to include the cost of time into the transportation costs, and thus into the cost of imported or exported food. Time increases the sailing costs for crew and auxiliary power, amount of refrigeration costs, and in the case in which it affects the type of cargo that can be transported, the price and revenue of the cargo. In the case of a speed reduction or slow steaming of ships, it may be expected that this cost would increase disproportionately for countries with longer distances to markets.

Certain national studies, like the one by the Centre for International Economy (Argentina) (CIE 2020) have shown that any short-term GHG reduction measure

Table 2. *List of the literature, considering slow steaming within eight criteria*

Geographic remoteness of and connectivity to main markets	Cargo value and type	Transport dependency	Transport cost	Food security	Disaster response	Cost-effectiveness	Socio-economic progress and development
Hummel's and Schaur (2012)	Hämäläinen (2014)	APEC (2019)	Zanne et al. (2013)	Carson (2016)	Psaraftis and Kontovas (2020)	Lindstad et al. (2012)	UNCTAD (2014)
Maloni et al. (2013)	Psaraftis and Kontovas (2014)	OECD (2014)	Psaraftis and Kontovas (2010)	Karampampa (2014)	UNCTAD (2014)	Faber et al. (2012)	Sanchez et al. (2003)
Wilmsmeier and Sanchez (2009)	Jivén et al. (2020)	Zanne et al. (2013)	Dagkinis and Nikitakos (2015)	Soyer, Tetten-born (2016)	Mimura (2013)	Finnsgård et al. (2020)	Wilmsmeier and Sanchez (2009)
Centre for International Economy (Argentina) (2020)	CEI (2019)	Wilmsmeier and Sanchez (2009)	Cepeda et al. (2017)		OECD (2014)	Zanne et al. (2013)	Newell et al. (2015)
Ferrari et al., (2015)	Karampampa (2014)	Holland et al. (2014)	Zhao et al. (2020)		Holland et al. (2014)	Cepeda et al (2017)	Jivén, et al. (2020)
APEC (2019)	Soyer, Tetten-born, (2016)	Mallidis et al. (2018)	Gurning et al. (2017)		Robinson (2020)	Mander (2017)	Doelle and Chircop (2019)
	Kontovas and Psaraftis (2011a)	UNCTAD (2014)	Wu (2020)			Centre for International Economy (Argentina) (2020)	
	Wiesmann (2010)	Yin et al. (2014)	Armstrong (2013)			Centre for International Economy (Argentina) (2021)	
	Wilmsmeier and Sanchez (2009)	Psaraftis and Zis, (2020)	Notteboom et al. (2021)			MEPC 62/INF.7 (2011)	
	Zanne et al. (2013)	Hummels et al. (2012)	Pradana et al. (2020)				
	Mander (2017)	Psaraftis and Kontovas (2010)	Elzarka & Morsi (2014)				
	UNCTAD (2014)	Oliveira (2014)	Carson (2016)				
		Mundaca et al. (2021)	Finnsgård et al. (2020)				
		Psaraftis and Kontovas (2020)	Cepeda & Caprace (2015)				
		Robinson (2020)	Balcombe et al. (2019)				
			Hämäläinen (2014)				
			Valentine et al (2013)				
			Centre for International Economy (Argentina) (2020)				
			Kontovas and Psaraftis (2011b)				

approved by the IMO would most probably result in increased costs of export for Argentine suppliers and reduced competitiveness of Argentina's foreign trade. These increased costs will be higher for longer transport distances and disproportionately distributed amongst countries as they largely depend on their distance from the main markets with Argentinian ports being more distant to China or the EU than US, Chinese or EU ports are amongst each other.

The impact of slow steaming on maritime services and inter-port competition focusing on the major shipping services and the effects on service patterns between Asia and Europe- was highlighted in a study by Ferrari, et al. (2015). According to their findings, slow steaming can lead to the offer of differentiated shipping services, combining fast and direct services among main hubs and cheaper and slower services calling on small ports.

Meanwhile, the APEC report (2019) investigated the economic and environmental impact of slow steaming for Intra-APEC long distant economies, as well as the parameters that need to be considered when evaluating slow steaming across varied ship types, fleet, distance, and cargo from a shipper's/trade point of view. The document highlighted that Intra-APEC long distance trade is restricted to sea, air or combination of both and analysed the nine long-distance Intra-APEC trade flows, i.e., three dry bulk trade routes and six containerized cargo trades with various values and types of cargoes. According to this report, slow steaming is more effective for ships with higher designed speeds, such as container ships and vehicle carriers. This document, through the analysis of various scenarios, clearly concluded that emission reductions will be eroded for longer distances and super slow steaming might lead to the need of additional ships.

3.2. Cargo value and type

Cargo value and type in relation with the speed reduction measure as a modality of reducing GHG emissions from shipping has been widely discussed in literature, with varying conclusions. Speed reduction can be beneficial for shippers when bunker prices are high and market rates are low, as well as when there is fleet over-capacity due to decreasing demand (Kontovas and Psaraftis 2011a). Hämäläinen (2014) considered that speed reduction and freight shipping could economically lead to a positive effect to shippers and cargo owners in the new fuel situation from 2015 onwards. This analysis has pointed out that there are quite vast variations between export markets due to different supply chain environments. Psaraftis and Kontovas (2014) highlighted that both fuel price and freight rates, which depend on the market conditions, have a strong effect on determining the optimum sailing speed of a vessel. Meanwhile, the effects for a specific shipping company depend on various factors such as shipping segment, geographic market, modal competition and the design of the service, and can vary from minor to significant (Jiven et al. 2020).

Speed reduction may reduce short-term operating costs and this could have an impact on the final customers (Carson 2016). Additionally, this issue could have a potential significant implication for shipping customers, particularly those shipping perishable goods such as food (Karampampa 2014; Soyer and Tettenborn 2016). This implies that the increase in transport costs due to speed reduction will also vary depending on the commodity type and on whether this cost reduction can be passed-through to the producers or other actors of the supply chain. As an example, in the case study of soybean meal exports to the Netherlands, 11 – 41% of the cost reductions were not passed through to the other actors; however, a reduction of 5 – 23% of the costs were passed through to the other actors (CEI 2019). There is also a relation between slow steam measure and transportation time (Wiesmann 2010). In fact, cargo owners have to accept that the transportation time of their goods will be increased slightly and at the same time, the carriers need to adapt their trade schedule. Wilmsmeier and Sanchez (2009), also highlighted the importance of including the cost of time into the transportation costs, and thus into the cost of imported or exported food.

Although many studies show that the overall benefits of slow steaming overweigh the costs, shippers will have their individual perspectives on the practice, which may vary with type and volume of the cargo, as well as the availability of credit facilities (Zanne et al. 2013). Additionally, according to Mander (2017), slow-steaming may benefit carriers in reducing fuel costs, but this benefit may not be passed on to shippers who may suffer from increased transit times of the goods, leading to unequally distributed advantages and disadvantages. The risk of increased transit times to the quality of perishable goods is also another important factor that must be considered, especially for SIDS and LDCs in terms of global markets, such as costs and connectivity issues, as well as affecting the reliability of transport and logistics services and impact on cargo value (UNCTAD 2014).

3.3. Transport dependency

Although the majority of studies addressed the transport dependency criterion, less attention has been paid to the impact of slow steaming on transport dependency for SIDS and LDCS. SIDS regroup a collection of countries that are diverse in many aspects, including in terms of their geographical location and respective levels of development. Despite some differences in the profile, structure and flows of their trade, SIDS share a number of common features from an international transport perspective: geographic remoteness from their main trade partners; limited volumes of trade; trade imbalances stemming from a heavy reliance on imports; and low volumes of exports highly concentrated in a few products (OECD 2014; Psaraftis and Zis 2021). As highly open economies, most SIDS are particularly dependent on their foreign trade and suffer from a strong exposure to external variations, including global or regional financial and economic crises. Also, due to their geographical location in areas of strong weather and seismic events, many SIDS find themselves amongst the most vulnerable territories in terms of exposure to natural hazards and foreseeable impacts of climate change (Robinson 2020). Both economic and environmental risks have significant bearings on SIDS' transport systems in terms of reliability and costly operation. UNCTAD (2014), highlighted some of the related obstacles faced by transport services connecting

SIDS to global markets, such as costs and connectivity issues, transport dependency, as well as disruptive weather related events affecting the reliability of transport and logistics services. Transport costs of SIDS trade are comparatively high because small volumes of trade have to travel long and indirect routes to reach distant markets.

Taking the above into consideration, Holland et al. (2014), provided context for an emerging discussion on the relationships between global CO_2 and other pollutant emissions, in particular the contributions from global shipping, and the sea transport scenario for Pacific Island countries (PICs). The Pacific is the most dependent region on imported fuels; a major inhibitor of sustainable development for PICs. Although most operators in PICs already employ slow steaming, considering that fuel costs are the major component of shipping operators' costs one of the options considered to reduce this cost is employing further slow steaming. This effort highlighted the problem of missing data and records in order to perform accurate impact analysis of the potential measures in order to mitigate the effect of GHG emissions in PICs. Sea transport is a critical need; the region's transport issues are unique-small and vulnerable economies, long distances, and old ships of poor standard, entirely fossil fuel dependent and minute contribution to global emissions.

Additionally, an APEC report (2019), investigated the results of the economic and environmental impacts of slow steaming for Intra-APEC long distant economies, as well as the parameters that need to be considered when evaluating slow steaming across varied ship types (two classes of container and bulk ships), fleet, distance, and cargo from a shipper's/trade point of view. That document highlighted that Intra-APEC long distance trade is restricted to sea, air or combination of both. The document analyses the nine long-distance Intra-APEC trade flows, i.e. three dry bulk trade routes and six containerized cargo trades with various values and types of cargoes. Depending on the commodity value and type, the study found slow steaming to have a different impact. For low valuable and non-perishable products, the study found the delay due to slow steaming to be minimal. However, for high value perishable goods or fast-moving consumer goods the study found the impact due to the delay caused by slow steaming to be considerable and that it may result in a shift to higher carbon intensive transport modes, such as air freight. Moreover, slow steaming is more effective for ships with higher designed speeds such as container ships, vehicle carriers. The document through analysis of various scenarios showed that in longer distances emission reductions will be eroded, as well as the implementation of super slow steaming might lead to requiring additional ships to ensure the service is maintained.

A certain number of the sourced articles examined the impact of slow steaming on transport dependency through analyzing the impact of slow steaming on cost. Zanne et al. (2013), highlighted that transport dependency is amalgamating with transport cost and Wilmsmeier and Sanchez (2009), conducted an analysis of the influence of transport, namely freight rates on food prices. The study carried out a regression analysis on how transport cost is affected by various factors including the maritime transport dependency and centrality of a country in the shipping liner network, allowing to investigate the impact of centrality on transport costs. This analysis showed that food prices, which are correlated with transport costs are higher in more geographically remote areas (which has more dependency to the maritime transportation), not only by geographical distance, but also by their position in global shipping networks. Other studies showed that slow steaming results to voyage cost reduction and it is not always correlated with transport distance with the largest delays occurring for the shippers on the last segment of the voyage (Mallidis et al. 2018). Additionally, a study on container ships sailing on the East Asia-North Europe trade route showed that both fuel price and a price on carbon lead to a reduction in the optimum sailing speed. The authors concluded that both an increase in fuel price or an increase in the price for CO_2 emissions should cause ship owners to reduce their sailing speed, and that this was correlated with a positive environmental impact in terms of GHG emissions reduction (Yin et al. 2014).

Hummels et al. (2012), estimated that each day in transit is equivalent to an ad-valorem tariff of 0.6 to 2.3 percent and that the most time-sensitive trade flows are those involving parts and components trade. These results suggest a link between sharp declines in the price of air shipping and rapid growth in trade as well as growth in world-wide fragmentation of production. Psaraftis and Kontovas (2010), analysed the cost of shipping under varying conditions of charter rates and ship speed. The study built a model to consider the modal split between shipping and one land-based transportation route. It was found that according to the model used, if the speed of ships was reduced at constant freight rates, then more cargo would be shifted to the land-route. Meanwhile, number of papers discussed the transport dependency through the possibility of modal shift in case of slow steaming and its negative externalities. As an example, slow steaming may encourage some cargoes to shift to other, faster modes of transport, including road, rail or air (Psaraftis and Zis 2021), which can lead to more air emissions to the environment as the case of modal shift to air freight is 47 times more carbon intensive than a container ship (APEC 2019).

Other studies considered the determinates of freights on the main shipping route. As an example, Oliveira (2014), analyzed the determinants of freight rates on main shipping routes linking Europe to the rest of the world. According to the study freight rates across routes can vary due to specific characteristics of the routes in terms of fuel cost, amount of transshipment, trade volume, connectivity and port infrastructure. Major routes have an advantage in terms of fuel cost, not only due to maritime distance, but also because of the deployment of the most efficient vessels on these routes. The amount of transshipment also affects the freight rates that tend to be higher for routes with low trade volume. In addition, setting rules for carbon reduction in shipping is another factor that is considered to be an effective measure in setting freight rates. Mundaca et al. (2021), highlighted the potential impacts of carbon taxation of international transport fuels on CO_2 emissions and trade activity, focusing on maritime transport. According to the findings, the bunker-price elasticities of products range from -0.003 to -0.42 and the largest impacts of carbon taxes on maritime trade are on the products with the lowest sales values in relation to their weight. A global tax of $40 per ton CO_2 tax would reduce carbon emissions by about 7% for the heaviest traded products transported by sea with the greatest CO_2 emission reductions being for products with particularly low value-toweight ratios such as fossil fuels and ores. Their conclusions suggest that an increase in the bunker fuel price would lead to substantial reductions per ton- km for internationally traded products reducing at the same time the bunker fuel consumption and carbon emissions from international shipping.

3.4. Transport cost

The majority of the peer-reviewed papers examined the impact of slow steaming on transport costs, with many of them analysing the direct impact on vessels' operational costs and others following a more holistic approach and discussing how slow steaming would affect the whole supply chain and the various actors involved. The important benefits of applying slow steaming in reducing fuel consumption and, therefore, fuel costs and harmful emissions were highlighted in the findings of Zanne et al. (2013). Psaraftis and Kontovas (2010) and Dagkinis and Nikitakos (2015), also analysed the cost of shipping under varying conditions of charter rates and ship speed and concluded that slower speed was the preferred option in terms of costs if the charter rates were low, whereas higher speeds were preferred at higher charter rates. Cepeda et al. (2017), also conducted sensitivity analyses with respect to bunker prices and the costs of new sheets to supplement the fleet, which showed that cost reductions favoured the use of slow steaming if bunker prices are high, and favoured operation at original speed (assuming constant transport work) if the capital expense of new vessels was high.

Zhao et al. (2020), carried out an analysis of the loss aversion mechanism employed by decision makers at shipping companies in order to determine the optimal sailing speed and recommended that shipping companies should reduce sailing speed under conditions of high fuel prices, low compensation costs required to be paid to customers for late deliveries, and when there were few cargoes onboard the vessels. Gurning et al. (2017), also investigated which ship speed is most optimal for shipping companies by considering technical and operational, financial and also environmental aspects and supported that the ship's speed is the most important factor in ship's operational cost and revenue, and operating in optimum speed could help the ship-owners save costs. The analysis recommended super slow steaming as the optimal speed for container vessels. Extra slow steaming and slow steaming alternatives are placed in second and third ranks, respectively and full speed is placed as the last optimal speed for container vessels. According to Wu (2020), carriers will reduce ship speed if the saved fuel cost outweighs the incurred capital and operating costs, and slow steaming can save fuel consumption for carriers even if additional ships may be required to maintain sailing frequency. Moreover, it was highlighted that the optimal speed will be increased by the ship size, and if the carriers intend to deploy slow steaming the optimal strategy is to call extra ports with extra ships.

Armstrong (2013), found slow-steaming to be the most effective measure for the reduction of fuel consumption and suggested that, even though the implementation of slow-steaming was the result of commercial optimisation in the light of the economic recession of 2008-2009, it would be prudent to continue the practice of slowsteaming even when the world economy recovers, in order to maintain the significant fuel efficiency benefits in the interest of reducing GHG emissions. Notteboom et al. (2021), investigated and compared the temporal and spatial sequences of the supply and demand shocks of COVID-19 and the 2008-2009 financial crisis on container ports and the container shipping industry. According to their findings, adaptation mechanisms to these shocks-such as slow steaming-have not been applied at the same extent during the financial crisis and COVID-19. More specifically, slow steaming that was extensively used as a cost-saving strategy during the financial crisis of 2008-2009 was not a part of the capacity management efforts during COVID-19, as this strategy was already in place and there was little room left for further speed reductions.

Pradana et al. (2020), used analytical relations to calculate the shipping costs for cargo ships operating in Indonesia (those that connect the more remote parts of eastern Indonesia with western Indonesia). Assumptions for ship speed were made and divided into three scenarios: In the first scenario the shipping costs were calculated using real voyage data made available by the Ministry of Transportation of the Republic of Indonesia. In the second scenario, the ship speed was assumed to be reduced based on reports about the weather conditions (wind and waves) using an assumption about ship speed loss as a function of weather. In the third scenario the ship speed was reduced further, not only as a result of the weather conditions, but also as a result of a further reduction of speed of either 10% or 12%. The results showed that the lowest transport costs were calculated for the third scenario, which simultaneously implemented a reduction of ship speed due to weather and slow steaming. For two out of the three shipping routes used in this investigation, the lowest transport cost was achieved for the 10% speed reduction in the third scenario.

A number of studies analysed the impact of slow steaming on the different actors of the supply. According to Elzarka and Morsi (2014) the impact of slow steaming on cost is undeniable, but supply chain lead times become longer and alternative supply sourcing options need to be examined (e.g. near-shoring or on-shoring). Carson (2016), stated that, even though speed reduction has reduced short-term operating costs for ship operators, shippers have not benefited from slow steaming to the same extent as carriers have not passed on savings from reduced shipping costs via lower freight rates. Similar findings come from the study of Finnsgård et al. (2020); the authors conducted a case study on six Swedish companies in various industries using shipping to transport their goods and focused on the impact of slow-steaming on the shipper, and not on the ship owner. The study noted that in principle there is a negative abatement cost to GHG abatement from slow-steaming. According to their findings, all companies reported that they felt increased transit times associated with slow-steaming, with only one of the companies receiving a recognisable reduction in transport cost, as a result of the slow-steaming practices. This means that all the cost-reduction due to this practice remained with the carrier.

Cepeda and Caprace (2015), assessed in particular whether the reduced transportation work due to slow-steaming or ultra-slow-steaming would still be profitable although new ships would need to be purchased to restore the transportation work to its original value. It was found that both slow-steaming and ultra-slow steaming were more profitable than the current practice, and that costs for purchasing new vessels were more than offset. Zanne et al. (2013), elaborated on the GHG and transport cost reduction benefits of slow steaming in the wake of the economic recession of 2008-2009 and concluded that slow-steaming is the most cost-effective measure of reducing fuel consumption and air emissions, but shippers will have their individual perspectives on the practice, which may vary with type and volume of the cargo, as well as the availability of credit facilities. Balcombe et al. (2019) found out that the average GHG savings due to slow steaming are around 19%, if the addition of further vessels to make up for the lost transport work is considered, and underlined that, while slow steaming results in considerable fuel savings, the cost reduction is not always felt by the cargo owner and slow-steaming practices may require some form of regulation. Hämäläinen (2014), examined how slow steaming in freight shipping can bring economically positive effects to shippers and cargo owners in the new fuel situation from 2015 onwards. According to this analysis, slow steaming has been used to tackle the negative economic impacts of the Sulphur Directive in the short sea shipping routes, but positive impacts of slow steaming vary from market to market. Valentine et al. (2013), also suggested that slow steaming consists of one of the measures adopted by the shipping industry as a response to rising fuel costs.

The 2020 report of the Centre for International Economy (Argentina) (2020) provides an analysis of the economic impact of energy efficiency measures in shipping on the cost of Argentine exports. According to this analysis, speed reductions of 10 – 30% could reduce actual cost of transportation, even if additional ships would be needed to maintain a constant transportation rate. The study highlights the importance of this costs reduction to be passed-through to the producers; for example, for the case of soybean exports to China, producers would face a cost increase by 11 – 37% or an actual reduction in their costs by 6 – 26% depending on whether the transport costs reductions would be passed through.

Kontovas and Psaraftis (2011), investigated the scenario of reducing emissions along the maritime intermodal container chain and explored the potential effects of speed reduction of container vessels on reducing emissions and fuel consumption. They investigated possible ways to decrease time in port since slow steaming increases the time of vessels at sea focusing on the potential to reduce berthing and waiting times at container ports to keep the total turnaround time constant. According to their findings, speed reduction can be beneficial for shippers when bunker prices are high and market rates are low and when there is fleet overcapacity due to decreasing demand. Concerning the cost-effectiveness of speed reduction, the authors pointed out that it is an overall more profitable option for the operators depending on the additional costs of deploying the extra vessels. For the charterers, speed reduction relates to increased in-transit inventory costs that are proportional to the value of the cargo.

3.5. Food security

A number of studies have identified slow-steaming as having a potential negative impact on food prices and food security. Some articles argued that slow steaming may have significant implications for the shipping customers of perishable goods such as food (Karampampa 2014; Carson 2016; Soyer and Tettenborn 2016) and Wilmsmeier and Sanchez (2009), conducted an analysis of the influence of transport, namely freight rates on food prices. In the introduction freight rates of the first part of this millennium leading up to the start of the economic crisis in the last quarter of 2008 are compared with food prices, indicating considerable correlation. The study notes that since the food importer bears the price of the transport costs, increased transport costs would be expected to have a considerable economic and societal impact on the costs and consumption of food. People with low incomes would be particularly vulnerable to an increase in food costs deriving from transport cost, and may thus affect food security. This analysis shows that food prices, which are correlated with transport costs are higher in more geographically remote areas, not only by geographical distance, but also by their position in global shipping networks, for example if they are situated at the start or end of shipping lines. Another important aspect discussed in this article is that it is necessary to include the cost of time into the transportation costs, and thus unto the cost of imported or exported food. Time increases the sailing costs for crew and auxiliary power, amount of refrigeration costs, and in the case in which it affects the type of cargo that can be transported, the price and revenue of the cargo. In the case of a speed reduction or slow steaming of ships, it may be expected that this cost would increase disproportionately for countries with longer distances to markets.

3.6. Disaster response

SIDS are heavily dependent on marine transportation and face common economic and development challenges. Geographical remoteness, the higher shipping costs associated with the low connectivity to main markets, the need for transhipment, and the transport dependency particularly during emergencies are some of the main concerns of SIDS (Psaraftis and Zis 2021).

Most SIDS are vulnerable to natural hazards (UNCTAD 2014), and like the other developing countries climate change and population growth is increasing the vulnerability of SIDS to disaster risk (Mimura 2013). On one hand, the larger exposure of SIDS to the impacts of climate change and natural disaster and on the other hand lack of ability of SIDS to respond, make SIDS among the most vulnerable countries in the world (OECD 2014). Connectivity of SIDS with the global community and international trade is crucial, to fulfil the essential needs of SIDS. Shipping is the "lifeline" and plays an important role for SIDS in essential connectivity and disaster response. Although the topic is crucial for SIDS, there is a lack of study, and neither papers nor case studies provide any analytical information on the extent of the importance of maritime transportation and the negative impacts of the implementation of slow steaming in relation to response in disasters.

As open and small economies, SIDS are also vulnerable to global economic and financial shocks. OECD (2014), highlighted that short-term disaster response must be linked to long-term financial support for resilience building and financing, and Holland et al. (2014), provided context for an emerging discussion on the relationships between global CO_2 and other pollutant emissions in particular the contributions from global shipping, and the sea transport scenario for PICs. The Pacific is the region most dependent on imported fuels, which is a major inhibitor of sustainable development for PICs. The paper highlighted that sea transport is a critical need and the region's transport issues are unique, small and vulnerable economies, long distances, and old ships of a poor standard, entirely fossil-fuel dependent and minute contribution to global emissions. Robinson (2020), highlighted that although SIDS are among the least responsible of all nations for climate change, they are likely to suffer strongly from its adverse effects and could in some cases even become uninhabitable. Ninety per cent of the SIDS are in the tropics, and almost all SIDS depend heavily on fossil fuels. The use of fossil fuels includes not only power production and the desalination of water, but also transport, including tourists as well as the transfer of goods.

3.7. Cost-effectiveness

A rather intermediate number of studies discussed the cost-effectiveness of vessel speed reduction as a means of reducing GHG emissions from shipping with varying conclusions. These studies found that speed reduction was a cost-effective means of reducing GHG emissions, in that it comes at no economic cost, or even at a negative GHG abatement cost. Part of this group of studies was document MEPC 62/INF.7 (2011) which concluded that a 10% or 20% speed reduction would both have negative abatement costs, but that a 10% speed reduction was most cost-effective. Lindstad et al. (2012), concluded at a reduction of GHG emissions of up to 28% being possible at zero abatement costs. The same study also concluded that global GHG emissions could be reduced by 30% over this time span at zero abatement cost, by increasing the size of the vessels. This suggests that while slow-steaming was estimated to have a negative abatement cost, the abatement potential of using larger vessels was even larger. Additionally, Finnsgård et al. (2020) highlighted that in principle the abatement cost of slow-steaming is negative and if the average speed of vessels in 2007 would be reduced to 85%, benefits would outweigh the costs by 178 – 617 billion USD in the period up to 2050, depending on fuel prices (Faber et al. 2012).

However, a number of studies showed a more balanced picture. Zanne et al. (2013) highlighted that while studies showed that the overall benefits of slow steaming overweigh the costs, shippers will have their individual perspectives on the practice, which may vary with type and volume of the cargo, as well as the availability of credit facilities. The authors further mentioned the potential for energy efficiency in propulsion and auxiliary power to contribute to a reduction in fuel consumption, and also highlighted the possibilities of using renewable energy such as wind and solar power, battery electric hybrid propulsion systems and alternative fuels to replace part of the energy currently derived from fuels. The authors concluded that slow-steaming is not the only way of reducing fuel consumption and air emissions, but that it appeared to be the most costeffective measure.

Cepeda et al. (2017) showed that slow steaming and ultra-slow steaming reduced CO_2 emissions per transport work by 51% and 85% respectively, whilst reducing at the same time SOx emissions by 43% and 78% respectively and operating costs of the fleet by 1% and 34% respectively. The fleet was supplemented with new ships in order to maintain the transport work at more or less a constant level ($\pm 7\%$). The study also conducted sensitivity analyses with respect to bunker prices and the costs of new ships to supplement the fleet, which showed that cost reductions favoured the use of slow steaming if bunker prices were high, and favoured operation at original speed (assuming constant transport work) if the capital expense of new vessels was high.

A number of studies found that slow-steaming could have severe economic costs. Mander (2017), highlighted the fact that slow-steaming may have a wider impact on supply-chains beyond shipping, with businesses operating lean 'just-in time' supply chains most affected. Slow-steaming may benefit carriers in reducing fuel costs, but this benefit may not be passed on to shippers who may suffer from increased transit times of the goods, leading to unequally distributed advantages and disadvantages. The risk of increased transit times to the quality of perishable goods is also mentioned. Slowsteaming requires coordination of the larger logistics network in order to ensure sufficient warehouse space and coordination with other modes of transport. The author also mentioned the impact of slow-steaming on ship crews, increasing their transit times and potentially requiring potential changes in duties onboard.

The 2020 report by the Centre for International Economy (Argentina) (2020), provided an analysis of the economic impact of energy efficiency measures in shipping on the cost of Argentine exports. Three energy efficiency measures were considered: 1. operational measures such as speed optimisation and speed reduction, as well as improvement of port operations, 2. technical measures such as energy efficiency measures and the use of low-carbon alternative fuels, and 3. the use of market-based measures such as negotiable emissions permits and emission taxes. Two case-studies using soybean product exports to China and the Netherlands were used to evaluate the costs of the energy efficiency measures. The study observed that having to pay for CO_2 emissions would result in a $10 - 36\%$ increase in transport costs for the case of Argentine soybean exports. The study also observed that speed reductions of $10 - 30\%$ could reduce actual cost of transportation, even if it was considered that additional ships would be needed to maintain a constant transportation rate. Crucially, the study noted that it depends on whether this cost reduction can be passed-through to the producers, on whether it would increase their cost by between 11 and 37% or in fact reduce their costs by $6 - 26\%$ for the case of soybean exports to China. In the case study of soybean meal exports to the Netherlands, similar values were calculated: An increase of $11 - 41\%$ in case the costs reductions were not passed through and a reduction of $5 - 23\%$ if the costs were passed through. Finally, it was noted that there exists a risk that increases in transport costs could result in agricultural produce suppliers from other countries may step in and supply the markets instead, but that this depends on how much of the cost-savings due to fuel reduction are passedon to the charterers.

Furthermore, the 2021 paper by the Centre for International Economy (Argentina) (2021), argues that GHG reduction measures approved by the IMO would result in increased costs of export for Argentine suppliers with high probability and this would result in a reduced competitiveness of Argentina's foreign trade. It was also argued that these costs would be higher for longer transport distances, and that therefore increases in costs would be disproportionately distributed amongst countries, and depend on their distance from the main markets. The paper aims to provide preliminary statistical data for analysing the impact of the short-term GHG reduction measure on trade. First, data was provided on the fraction of trade executed by shipping, which in the case of Argentina is much higher at 75% than the United States (46%) or the EU (35%). It was also argued that Argentine ports have longer distances to China or the EU, than equivalent US, Chinese or EU ports have amongst each other. Data was also provided on the industrial sectors on which Argentina depends for exports. Argentina's exports rely predominantly on agriculture food production, fish and minerals (which are predominantly transported by ship), with the share of industrial products being relatively low at 22% of value. Finally, data was provided on the chilled and frozen meat exports, on which Argentina strongly relies. It was shown that chilled meat exports provide higher revenues, because of the higher market price for chilled meats. If shipping speeds are reduced, Argentina will be impacted by losing markets for chilled meat, and thus loose revenue. It may be able to make up trade losses to some extent by exporting frozen meat instead, if possible, but export losses would persist due to the lower market prices for frozen meat. These negative economic impacts may need to be considered in the cost-effectiveness of the measures.

3.8. Socio-economic progress and development

Socioeconomic progress and development can be defined as the identification and inclusion of factors determining the rate of the socioeconomic progress of a country under definite conditions of its development process (Gilani and Hoseinzadeh 2021; Khodayar Sahebi et al. 2021; Hoseinzadeh and Garcia 2022; Vakili et al. 2022b). The change to a longterm slow steaming scenario need, however, a number of considerations. Most SIDS are particularly dependent on their foreign trade and suffer from a strong exposure to external variations, including global or regional financial and economic crises. Both economic and environmental risks have significant bearings on their transport systems in terms of reliability, costly operation and connectivity issues (UCTAD 2014). According to Sanchez et al. (2003), on average, 8.6% of the value of merchandise imported by the countries of Latin America and the Caribbean is spent on freight and insurance costs relating to their international carriage; this figure is 40% higher than the world average. Major differences persist within the region, with the Caribbean economies recording the highest indices.

Wilmsmeier and Sanchez (2009), highlighted in their analysis that food prices, which are correlated with transport costs, are higher in more geographically remote areas, not only by geographical distance, but also by their position in global shipping networks, for example if they are situated at the start or end of shipping lines. The effects for a specific shipping company depend on various factors, such as shipping segment, geographic market, modal competition and the design of the service, and can be varied from minor to significant (Jiven et al. 2020). In addition, according to Doelle and Chircop (2019) there is a necessity of further assessment for the overall effect of speed management, especially regarding safe navigation and economic development for SIDs and LDCs. As an example, Pacific islands due to their specific characteristics, such as severe dependency on fossil fuel, long routes, minute economies, imbalance in import and export, financial barriers, and high infrastructure cost, face greater challenges in achieving long-term sustainable shipping (Newell et al. 2015). Table 3 summarises our findings on the impact of slow steaming on countries, and most specifically, the LDCs and the SIDS.

4. DISCUSSION

From a global perspective, the impact of slow steaming as an emission reduction strategy will vary from market to market and geographic regions (Valentine et al. 2013; Hämäläinen 2014). As highlighted in some studies, when implementing slow steaming in shipping, it is crucial to maintain a balance between reducing emissions and extending transit times, while taking into account customer tolerance for longer transit times (APEC 2019). If the balance is disturbed and customer tolerance is exceeded, modal shifts to modes with significantly higher carbon intensity, such as air cargo, or, in the worst case, the potential loss of trade for highly time-sensitive cargoes may occur. In line with this, the current sequences of the supply and demand shocks due to the COVID-19, which have different origins and impacts compared to the 2008-2009 financial crisis, should have been taken into consideration for further projection analysis. Slow steaming, adopted during the financial crisis of 2008-2009 in an attempt to absorb excess capacity, could not be used as a tool during the pandemic as the practice of slow steaming was largely in place as the response of shipping lines to reduced shipping volumes caused by the global financial crisis of 2008-2009 (Armstrong 2013). Additionally, slow steaming is not a measure that can be applied universally and is contextual to the technology the industry is operating with and market conditions with the newbuilding and retrofitted vessels already operating at lower speeds.

Developing, SIDS, and LDCs vary in many ways, including in terms of their geographic location, their respective level of development, and their exposure to natural disasters and the likely impacts of climate change (UNCTAD 2014), but they are all strongly affected by the negative impacts of climate change. The Pacific region is the most dependent on imported fuels and since fuel costs are the major component of ship operators' costs, they are a major constraint on the sustainable development of PICs (APEC 2019). Therefore, slow steaming is one of the options to reduce these costs and currently most operators in the region are already using it whenever possible.

Table 3. *Impact of slow steaming on countries based on the eight criteria*

Eight criteria	Impact of slow steaming on LDCs and SIDS
Geographic remoteness of and connectivity to main markets	Increased 'pipeline inventory' and overall transport costs for areas distant to main markets due to their geographical distance, but also their position in global shipping networks Increased export costs and reduced competitiveness of foreign trade for longer transport distances and disproportionate impact distribution for countries distant to the main markets Impact on inter-port competition focusing on the major shipping services and leading to the offer of differentiated shipping services, combining fast and direct services among main hubs and cheaper and slower services calling on small ports
Cargo value and type	The impact depends on various factors such as shipping segment, geographic market, modal competition and the design of the service, and can vary from minor to significant Potential significant implication for shipping customers, particularly those shipping perishable goods such as food Will also vary depending on the commodity type and on whether this cost reduction can be passed-through to the producers or other actors of the supply chain The risk of increased transit times to the quality of perishable goods, especially for SIDS and LDCs in terms of connectivity issues, as well as the reliability of transport and logistics services and impact on cargo value
Transport dependency	The SIDS region's transport issues are unique—small and vulnerable economies, long distances, and old ships of poor standard, entirely fossil fuel dependent and minute contribution to global emissions If the speed of ships was reduced at constant freight rates, then more cargo would be shifted to the land-route for short distance shipping routes Slow steaming may encourage some cargoes to shift to other, faster modes of transport, including road, rail or air, which can lead to more air emissions to the environment Freight rates across routes can vary due to specific characteristics of the routes in terms of fuel cost, amount of transshipment, trade volume, connectivity, and port infrastructure. Major routes have an advantage in terms of fuel cost, not only due to maritime distance, but also because of the deployment of the most efficient vessels on these routes. The amount of transshipment also affects the freight rates that tend to be higher for routes with low trade volume
Transport cost	Important benefits in reducing fuel consumption and, therefore, fuel costs and harmful emissions Shipping companies should reduce sailing speed under conditions of high fuel prices, low compensation costs required to be paid to customers for late deliveries, and when there were few cargoes onboard the vessels The impact of slow steaming on cost is undeniable, but supply chain lead times become longer and alternative supply sourcing options need to be examined (e.g. near-shoring or on-shoring) Even though speed reduction has reduced short-term operating costs for ship operators, shippers have not benefited from slow steaming to the same extent as carriers have not passed on savings from reduced shipping costs via lower freight rates Slow-steaming is the most cost-effective measure of reducing fuel consumption and air emissions, but shippers will have their individual perspectives on the practice, which may vary with type and volume of the cargo, as well as the availability of credit facilities An overall more profitable option for the operators depending on the additional costs of deploying the extra vessels. For the charterers, speed reduction relates to increased in-transit inventory costs that are proportional to the value of the cargo

(Cont.)

Table 1. *Cont.*

Eight criteria	Impact of slow steaming on LDCs and SIDS
Food security	Slow-steaming having a potential negative impact on food prices and food security May have significant implications for the shipping customers of perishable goods such as food People with low incomes would be particularly vulnerable to an increase in food costs deriving from transport cost and may thus affect food security Necessary to include the cost of time into the transportation costs, and thus unto the cost of imported or exported food. Time increases the sailing costs for crew and auxiliary power, amount of refrigeration costs, and in the case in which it affects the type of cargo that can be transported, the price and revenue of the cargo
Disaster response	SIDS are heavily dependent on marine transportation and consist small and vulnerable economies that face common economic and development challenges due to geographical remoteness, the higher shipping costs associated with the low connectivity to main markets, the need for transhipment, and the transport dependency particularly during emergencies
Socio-economic progress and development	Most SIDS are particularly dependent on their foreign trade and suffer from a strong exposure to external variations, including global or regional financial and economic crises. Both economic and environmental risks have significant bearings on their transport systems in terms of reliability, costly operation and connectivity issues Pacific islands due to their specific characteristics, such as severe dependency on fossil fuel, long routes, minute economies, imbalance in import and export, financial barriers, and high infrastructure cost, face greater challenges in achieving long-term sustainable shipping

In general, freight rates for SIDS and LDCs are high (De Oliveira 2014), which is due more to the monopoly control of the market by shippers than to connectivity with distance to major routes. Therefore, due to the lack of adequate policies and funding, some literature recommends considering direct and indirect compensation mechanisms to address the concerns of SIDS and LDCs. Almost all SIDS rely heavily on fossil fuels for transportation (APEC 2019) and due to barriers such as cost and connectivity issues, transportation services to connect SIDS and LDCs to global markets may be affected. Moreover, the reliability of transport and logistics services for SIDS and LDCs is unique and should be taken into account when assessing the impacts of proposed measures.

Taking into account the above and changes in the strategic behavior of market participants, further adjustment mechanisms, such as slow steaming, have been shown to lead to different outcomes and uncertainty, especially for SIDS, LDCs and developing countries. However, it would be useful to use historical data on the application of slow steaming, especially after the 2008 financial crisis, and consider a holistic, systematic, and transdisciplinary approach to assess the direct and indirect impacts of slow steaming on all maritime cluster stakeholders (Vakili et al. 2022a), especially for SIDS and LDCs.

As indicated in several studies, it is important that all actors in the supply chain adopt slow steaming (APEC 2019; CIE 2020; CIE 2021). If or when slow steaming is made mandatory, the windows of berths in ports, for example, will have to be reviewed and adapted to the new operational regime; i.e., no port fees should be charged for slow steaming. Slow steaming may also be economically problematic and practically difficult for small passenger ships and mixed passenger/cargo ships, which are a lifeline for connecting many SIDS, LDCs and microstates. Passenger ships are the only transportation option for the connections of many SIDS and LDCs. However, the impact of slow steaming on passenger ships, feeder and sub feeder transport was not adequately analyzed in the peer-reviewed papers, and the results of the impact assessments were mainly based on qualitative literature reviews, with no segmentation provided. In this context, a temporary exemption regime from slow steaming could be proposed for certain vessel types and sizes serving SIDS and LDCs, but it should be emphasized that further research is needed since SIDS are not a homogeneous group and their specific characteristics should be considered.

The research effort found that most studies focused primarily on the economic impact of the use of speed reduction by ship owners, without a proper quantitative impact assessment of the other relevant supply chain actors. As several authors suggested during the peer review, slow steaming is an important measure that improves economic efficiency and environmental sustainability, as it directly affects the amount of fuel used, which is directly proportional to fuel costs and emissions (Zanne et al. 2013). However, the economic benefit gained by the ship owner from slow steaming is not equally distributed among the different actors of the global supply chain (Mander 2017). This suggests that while slow steaming will reduce operational costs, it will simultaneously lead to longer lead times in the supply chain and that alternative supply options should be considered to overcome this challenge. For many SIDS, LDCs and even developing countries, the demand for maritime transport is inelastic. In this sense, it is questionable whether any cost savings will ultimately be passed on to shippers and states and to what extent.

Another assumption in most impact assessment studies is that overall transportation costs will decrease as a result of slow shipping and lower operating costs due to less fuel consumption (CEI 2019). However, speed reduction equates to less transportation work that may lead to the use of more or larger vessels (Lindstad et al. 2012). The impact of such a development on more air emissions, ports and overall logistics costs needs to be assessed, as a larger number of ships may lead to long waiting times at ports and higher inventory costs borne by shippers and states and inventory costs directly related to cargo value. This will make more sense for SIDS and LDCs that are geographically remote and lack good connectivity to key markets. In these states, logistics costs represent a significant portion of total transportation costs, with much transshipment required for goods to reach their final destination.

Another important factor in the impact assessment of slow steaming is cargo type. Cargo may be time-sensitive due to the nature of its industries, such as perishable goods, high-value goods, or high turnover consumer goods (FMCG), is strongly affected by changes in transportation time and will seek the option that meets the cargo's required delivery dates, which may lead to shifts to more carbon-intensive modes, such as air freight (Psaraftis and Kontovas 2010; APEC 2019; Psaraftis and Zis 2021). As highlighted in several studies, product characteristics and total transportation costs are important factors in the choice of transport mode; perishable, high-value products may prefer air transport over sea transport because the value of the product can absorb the typically higher air freight costs, and time-sensitive products, such as FMCG, are also more likely to choose air transport over sea transport to meet delivery deadlines in their supply chain. Perishable goods in particular can be very sensitive to speed reductions. Some authors suggested that the effect of reduced operating costs due to speed reductions should be examined in relation to the importance of transit time in trade goods. It is important to emphasize that transit time may be of secondary importance for bulk and simple industrial goods, but is critical for perishable goods (APEC 2019).

5. CONCLUSIONS

The IMO has adopted the EEXI and CII as short term measures for decarbonisation of the shipping industry and the relevant data gathering and reporting scheme is obligatory since January 2023. However, many existing ships do not comply with EEXI and, due to their age, cannot invest in other technologies to reduce air emissions. Given the current barrier, the application of slow steaming may be an propionate measure to meet EEXI and CII requirements.

This qualitative systematic literature review was carried out as part of the comprehensive impact assessment of the amendments to MARPOL Annex VI on the approved short-term measure to reduce the carbon intensity of international shipping as approved by MEPC 75. The aim of this literature review is to identify potential impacts of the approved short-term measure on States, including developing countries, in particular, LDCs and SIDS. This effort was focused on documents related to the technical and operational implementation of the approved short-term measure, and includes relevant literature identified therein, peer-reviewed articles in the open-literature, relevant grey literature documents (non-peer reviewed documents), and any documents identified by the Steering Committee and the IMO Secretariat. The approved short-term measure defined a goal of carbon intensity reduction, rather than prescribing a means of achieving it. This literature review focussed mainly on speed reduction, on which there is ample material available, and which may be implemented with relative simplicity and negligible initial cost.

The analysis of the literature was carried out in the context of the short-term GHG reduction measure; it revealed a stark variation in the frequency with which the open literature treated the eight impact criteria. The majority of studies addressed the impact of slow-steaming on transport dependency and transport costs of speed reduction. A high number of studies considered cargo value and type when discussing potential impacts. An intermediate number of studies discussed cost-effectiveness, socio-economic progress and development, and geographic remoteness and connectivity to main markets. Only a few studies addressed the issues of food security and disaster response.

Slow steaming, as a means of complying with carbon intensity reductions in the shortterm, was also found by some peer-reviewed literature studies to represent a key measure that reduces greenhouse gas emissions at high economic efficiency since it directly affects the quantity of fuel used that is directly proportional to fuel costs and emissions. On the other hand, it was a salient feature of the literature related to speed reduction, that economic benefits gained by the ship-owner were not necessarily equally distributed among the different actors of the global supply chain. It was also suggested in the literature that although slow steaming may be expected to reduce operational costs, it will at the same time result in longer supply chain lead times resulting in the need to consider alternative supply sourcing options to overcome this challenge.

The literature showed that there are certain features of transport services connecting SIDS and LDCs to global markets that are unique such as costs and connectivity issues, transport dependency as well as disruptive events affecting the reliability of transport and logistics services and need to be taken into consideration when assessing the impacts of proposed measures. Lack of a proper and up-to-date set of data and pertinent studies specifically focused on SIDS and LDCs, still constitutes a major shortcoming in order to perform a detailed and accurate impact assessment. Therefore, due to lack of data availability, most of the studies focussed on qualitative rather quantitative analysis.

Although slow steaming is expected to bring large economic and environmental benefits to the states whose economy depends heavily on international seaborne transport, an holistic and transdisciplinary approach needs to be adopted for conducting a comprehensive impact assessment. In this direction, further research on this topic should ideally consider employing methodologies and approaches to deal with conflicting objectives under trade off environment, such as from a simple Cost/Benefit Analysis to more advanced Multiple Criteria Decision Making tools.

Abbreviations

CII	Carbon intensity indicator
DCS	Data collection system
EEDI	Energy efficiency design index
EEXI	Energy efficiency existing ship index
GDP	Gross domestic product
GHG	Greenhouse gas
IMO	International maritime organization
LDCs	Least developed countries
MEPC	Marine environment protection
PICs	Pacific island countries
SEEMP	Ship energy efficiency management plan
SIDS	Small island development states

Acknowledgements

The authors would like to thank the reviewers and the journal's editor for their valuable comments, which have greatly improved the study. The authors are also grateful to ChinaMerchants Energy Shipping that support us in charging the processing of this work.

Author contributions

All authors read and approved the final manuscript.

REFERENCES

1. APEC (2019) Analysis of the impacts of slow steaming for distant economies, APEC Transportation Group, APEC Project: TPT 03 2018A, produced by Starcrest Consulting Group for the Asia-Pacific Economic Cooperation Secretariat, December.
2. Armstrong VN (2013) Vessel optimisation for low carbon shipping. Ocean Eng 73:195-207
3. Balcombe P, Brierley J, Lewis C, Skatvedt L, Speirs J, Hawkes A, Staffell I (2019) How to decarbonise international shipping: options for fuels, technologies and policies. Energy Convers Manag 182:72-88
4. Carson JK (2016) The implications of slow steaming for shipping customers
5. Centro de Economia Internacional (CEI) (2019) Greenhouse gas reduction measures in maritime transport and their impact on the cost of Argentine exports.. Greenhouse gas reduction measures in maritime transport and their impact on the cost of Argentine exports, https: / / cancilleria.gob.ar / es / cei / publicaciones
6. Centre for International Economy (2020) The potential impact of measures to reduce GHG emissions from ships on exports: a statistical preliminary approach Report
7. Centre for International Economy (2021) The potential impact of measures to reduce GHG emissions from ships on exports: a statistical preliminary approach Report
8. Cepeda MA, Assis LF, Marujo LG, Caprace JD (2017) Effects of slow steaming strategies on a ship fleet. Marine Syst Ocean Technol 12(3):178-186
9. Cepeda MAFS, Caprace JD (2015) SIMULATING ECONOMICAL IMPACTS OF SLOW & ULTRA SLOW STEAMING STRATEGIES ON A BULK CARRIER FLEET. In: Proceeding of the XXIX Congresso de Pesquisa e Ensino em Transportes (pp 1052 – 1062)
10. Dagkinis IOANNIS, Nikitakos NIKITAS (2015) Slow steaming options investigation using multi criteria decision analysis method. In: ECONSHIP 2015 Chios Greece.
11. De Oliveira GF (2014) Determinants of European freight rates: the role of market power and trade imbalance. Transp Res Part E Logist Transp Rev 62:23-33
12. Doelle M, Chircop A (2019) Decarbonizing international shipping: an appraisal of the IMO's Initial Strategy. Rev Eur Comp Int Environ Law 28(3):268-277
13. Elzarka S, Morsi M (2014) The supply chain perspective on slow steaming. In: International Forum on Shipping, Ports and Airports (IFSPA) 2014: Sustainable Development in Shipping and Transport LogisticsHong Kong Polytechnic University.
14. Faber J, Nelissen D, Hon G, Wang H, Tsimplis M (2012) Regulated slow steaming in maritime transport. An assessment of options, costs and benefits.
15. Ferrari C, Parola F, Tei A (2015) Determinants of slow steaming and implications on service patterns. Marit Policy Manag 42(7) : 636 – 652
16. Finnsgård C, Kalantari J, Roso V, Woxenius J (2020) The Shipper's perspective on slow steaming-Study of Six Swedish companies. Transp Policy 86:44-49
17. Fourth IMO GHG Study (2020) Final Rep. IMO doc. MEPC 75/7/15.

18. Gilani HA, Hoseinzadeh S (2021) Techno-economic study of compound parabolic collector in solar water heating system in the northern hemisphere. Appl Therm Eng 190:116756

19. Gurning RS, Busse W, Lubnan M (2017) Decision Making of Full Speed, Slow Steaming, Extra Slow Steaming and Super Slow Steaming using TOPSIS. Int J Marine Eng Innov Res 2(1)

20. Hämäläinen E (2014) Can slow steaming lower cost impacts of sulphur directive-shippers' perspective. World Rev Intermodal Transp Res 5(1):59-79

21. Holland E, Nuttall P, Newell A, Prasad B, Veitayaki J, Bola A, Kaitu'u J (2014) Connecting the dots: policy connections between Pacific Island shipping and global CO2 and pollutant emission reduction. Carbon Manag 5(1):93-105

22. Hoseinzadeh S, Garcia DA (2022) Techno-economic assessment of hybrid energy flexibility systems for islands' decarbonization: A case study in Italy. Sustain Energy Technol Assess 51:101929

23. Hummels D, Schaur G (2012) Time as a trade barrier (No. w17758). National Bureau of Economic Research

24. Jiven K, Lammgård C, Woxenius J, Fridell E (2020) Förstudie initierad av Lighthouse, E. Consequences of speed reductions for ships.

25. Karampampa IC (2014) THE IMPACT OF SLOW STEAMING ON SHIPPERS AND ON THEIR SUPPLY CHAINS: A WINDOW OF OPPORTUNITY FOR OTHER TRANSPORT MODES. Erasmus University Rotterdam.

26. Khodayar Sahebi H, Hoseinzadeh S, Ghadamian H, Ghasemi MH, Esmaeilion F, Garcia DA (2021) Techno-economic analysis and new design of a photovoltaic power plant by a direct radiation amplification system. Sustainability 13(20) : 11493

27. Kontovas CA, Psaraftis HN (2011a) The link between economy and environment in the post-crisis era: lessons learned from slow steaming. Int J Decis Sci Risk Manag. 3(3-4):311-326

28. Kontovas CA, Psaraftis HN (2011b) The link between economy and environment in the post-crisis era: lessons learned from slow steaming. Int J Decis Sci Risk Manag 3(3-4):311-326

29. Lindstad H, Asbjørnslett BE, Strømman AH (2012) The Importance of economies of scale for reductions in greenhouse gas emissions from shipping. Energy Policy 46:386-398

30. Mallidis I, lakovou E, Dekker R, Vlachos D (2018) The impact of slow steaming on the carriers' and shippers' costs: the case of a global logistics network. Transp Res Part E Logis Transp Rev 111:18-39

31. Maloni M, Paul JA, Gligor DM (2013) Slow steaming impacts on ocean carriers and shippers. Marit Econom Logist 15(2) : 151 – 171

32. Mander S (2017) Slow steaming and a new dawn for wind propulsion: a multi-level analysis of two low carbon shipping transitions. Mar Policy 75:210-216

33. Marine Environmental Protection Committee (MEPC) (2011) Reduction of GHG emissions from ships, Marginal abatement costs and cost effectiveness of energy efficiency measures. Submitted by (IMarEST). MEPC. 62/INF.7.

34. Marine Environmental Protection Committee (MEPC) (2018) INITIAL IMO STRATEGY ON REDUCTION OF GHG EMISSIONS FROM SHIPS. MEPC 72/17. ADD. 1

35. Marine Environmental Protection Committee (MEPC) (2019) PROCEDURE FOR ASSESSING IMPACTS ON STATES OF CANDIDATE MEASURES. MEPC. 1/ Circ. 885

36. Marine Environmental Protection Committee (MEPC) (2020) TERMS OF REFERENCE AND ARRANGEMENTS FOR THE CONDUCT OF A COMPREHENSIVE IMPACT ASSESSMENT OF THE SHORT-TERM MEASURE BEFORE MEPC 76. MEPC 75/18.

37. Mimura N (2013) Sea-level rise caused by climate change and its implications for society. Proc Jpn Acad Ser B 89(7) : 281 – 301

38. Mundaca G, Strand J, Young IR (2021) Carbon pricing of international transport fuels: Impacts on carbon emissions and trade activity. J Environ Econ Manag 110:102517

39. Newell A, Nuttall PM, Holland E (2015) Sustainable sea transport for the Pacific Islands: the obvious way forward. UN Sustainable Development Knowledge Platform

40. Notteboom T, Pallis T, Rodrigue JP (2021) Disruptions and resilience in global container shipping and ports: the COVID-19 pandemic versus the 2008-2009 financial crisis. Marit Econ Logist 23(2):179-210

41. OECD. Making Development Co-operation Work for Small Island Developing States (2014) Retrieved from: Making Development Co-operation Work for Small Island Developing States I en I OECD

42. Pradana MF, Hamdani MI, Noche B (2020) Shipping cost optimization on the Indonesian sea tollway due to weather. IOP Conf Ser Mater Sci Eng 909(1):012042

43. Psaraftis HN, Zis T (2021) Impact assessment of a mandatory operational goal-based short-term measure to reduce GHG emissions from ships: the LDC/SIDS case study. In: International Environmental Agreements: Politics, Law and Economics (pp 1-23).

44. Psaraftis HN, Kontovas CA (2010) Balancing the economic and environmental performance of maritime transportation. Transp Res Part D Transp Environ 15(8):458-462

45. Psaraftis HN, Kontovas CA (2014) Ship speed optimization: concepts, models and combined speed-routing scenarios Transp Res Part C Emerg Technol 44:52-69

46. Psaraftis HN, Kontovas CA (2020) Decarbonization of maritime transport: is there light at the end of the tunnel? Sustainability. https://doi.org/10.3390/su13010237

47. Robinson SA (2020) Climate change adaptation in SIDS: a systematic review of the literature pre and post the IPCC Fifth assessment report. Wiley Interdiscip Rev Clim Change 11(4):e653

48. Sánchez RJ, Hoffmann J, Micco A, Pizzolitto GV, Sgut M, Wilmsmeier G (2003) Port efficiency and international trade: port efficiency as a determinant of maritime transport costs. Marit Econ Logist 5(2):199-218

49. Soyer B, Tettenborn A (2016) Slow steaming clauses and international sales contracts: a successful marriage?. In: International Trade and Carriage of Goods (pp 55-74). Informa Law from Routledge.

50. UNCTAD (2014) Review of maritime transport. Chapter 6: Sustainable Freight Transport Development and Finance.

51. UNCTAD (2021) Review of maritime transport. Retrieved from https://unctad.org/webflyer/review-maritime-trans port-2021

52. Vakili SV, Ölçer Al, Schönborn A (2021) Identification of shipyard priorities in a multi-criteria decision-making environment through a Transdisciplinary energy management framework: a real case study for a Turkish shipyard. J Marine Sci Eng 9(10) : 1132
53. Vakili SV, Ballini F, Dalaklis D, Ölçer Al (2022a) A Conceptual transdisciplinary framework to overcome energy efficiency barriers in ship operation cycles to meet imo's initial green house gas strategy goals: case study for an iranian shipping company. Energies 15(6):2098
54. Vakili S, Schönborn A, Ölçer Al (2022b) Techno-economic feasibility of photovoltaic, wind and hybrid electrification systems for stand-alone and grid-connected shipyard electrification in Italy. J Clean Prod 366:132945
55. Vakili S, Ölçer Al, Schönborn A, Ballini F, Hoang AT (2022c) Energy-related clean and green framework for shipbuilding community towards zero-emissions: a strategic analysis from concept to case study. Int J Energy Res.
56. Valentine VF, Benamara H, Hoffmann J (2013) Maritime transport and international seaborne trade. Marit Policy Manag 40(3) : 226 – 242
57. Wiesmann A (2010) Slow steaming-a viable long-term option? Wartsila Techn J 2:49-55
58. Wilmsmeier G, Sanchez RJ (2009) The relevance of international transport costs on food prices: endogenous and exogenous effects. Res Transp Econ 25(1):56-66
59. Wu WM (2020) The optimal speed in container shipping: theory and empirical evidence. Transp Res Part E Logist Transp Rev 136:101903
60. Yin J, Fan L, Yang Z, Li KX (2014) Slow steaming of liner trade: its economic and environmental impacts. Marit Policy Manag 41(2):149-158
61. Zanne M, Počuča M, Bajec P (2013) Environmental and economic benefits of slow steaming. Trans Marit Sci 2(02) : 123 – 127
62. Zhao Y, Zhou J, Fan Y, Kuang H (2020) Sailing speed optimization model for slow steaming considering loss aversion mechanism. J Adv Transp.

INDEX